shoe.

SILVER

SILVER IN THE GOLDEN STATE

Images And Essays Celebrating The
History And Art

OF

Silver In California

ESSAYISTS

LESLIE GREENE BOWMAN	JOHN W. BURKE
DEBORAH COOPER	DONALD CHAPUT
ELLIOT EVANS	DONALD L. HARDESTY
EDGAR W. MORSE	GERALD W. R. WARD

Edgar W. Morse
EDITOR

L. Thomas Frye
INTRODUCTION

COLOR PLATES BY
Stephen Rahn

ASSOCIATE EDITOR
Thomas Curran

THE OAKLAND MUSEUM HISTORY DEPARTMENT

OAKLAND
1986

Contents

Introduction

IF IT WAS A QUEST for "Glory, Gold and God" that lured Spaniards north from Mexico into Alta California, their disappointment must have been keen. Things in California were not as they had been in South and Central America and Mexico, where the indigenous high cultures had revealed (and ultimately yielded) riches in gold and silver beyond the *Conquistadores'* wildest dreams; in California there simply was no *El Dorado* to be found. To the contrary, California was a sparsely settled frontier of scattered missions, presidios and pueblos, peopled by missionaries saving souls (but not always the lives) of the native peoples, and by soldiers protecting Spain's northwestern flanks against incursions from other European powers. Despite both the discovery of quicksilver (mercury) at New Almaden in Santa Clara county by Antonio Suñol and Luis Chaboya in 1824, and small, tantalizing, discoveries of gold in San Feliciano Cañon by Francisco López in 1842, the vast mineral wealth of California remained hidden from Hispanic eyes.

It was Americans, spurred both by President Polk and by the call of their own Manifest Destiny, who ultimately infiltrated and—with sheer force of numbers—overwhelmed Mexican California, claiming both the state and the secret she held. For the Americans, James Marshall's discovery of gold at Sutter's Mill in 1848 was as predictably providential as the Mexican cession of California itself that same year, which concluded both the Mexican War and *Californio* rule in the state.

As news of Marshall's discovery spread, the world rushed to America's *El Dorado*, and transformed California overnight from a remote frontier into the Golden State. And, frankly mixing metaphors, the pot of gold at the end of the California rainbow quickly proved itself to be silver-lined. Less known and less celebrated than the history of California gold, the history of California silver is the proverbial other side of the golden California coin.

In 1859, several gold miners—late of California placers—were working claims on the slopes of obscure Sun Mountain (eventually Mount Davidson) in Nevada. When they found that their gold pans, riffle-boxes, Long Toms, and sluices quickly became clogged with a "blue-looking stuff," several of their more enterprising number trekked across the Sierra Nevada Mountains to California to have the offending material assayed in Grass Valley and Nevada City. The "stuff" was found to be incredibly rich silver ore. Word got out; there promptly followed the "Rush to the Washoe" and the Big Bonanza—the mining of the famous Comstock Lode.

The exploitation of the Comstock forged an intimate and enduring link between the Nevada Territory and California, a link which, as Donald L. Hardesty points out, "... catapaulted San Francisco from a port town of middling status to the most important city on the West Coast of North America." Farther south, the silver mines of Cerro Gordo in Southern California contributed many millions of dollars towards the development of Los Angeles, as ton upon ton of silver-lead ingots from the "Fat Mountain" were hauled to the City of the Angels for transport to distant refineries.

But Silver mining and the financial empires it spawned are only part of our story. Comstock and Cerro Gordo silver effected the coronations of "Silver Kings" in San Francisco; it built mansions, luxurious hotels, and exclusive private clubs; and it created a new social elite, who demanded material possessions commensurate with their recently silver-lined pockets. With this ready-made clientele for their wares—and equally ready sources of available silver metal—silversmiths emerged in nineteenth century California who were both willing and able to produce elaborate, expensive objects for the nabobs, and, with help from new mass production technologies, wares for the emerging middle-class (eager to now emulate where formerly they could only envy). Table-wares, trophies, elaborate presentation pieces, souvenirs, harness, ship services, weapons, jewelry, *objets d'art*—and more—emerged from the hands and shops of California silversmiths. Judging by the proliferation of silverware in the 1860s and 1870s, California had arrived!

She had. And she hasn't looked back since.

For *Silver in the Golden State*, both this book and the exhibition of the same name which it accompanies, the Oakland Museum History Department owes substantial debts to many individuals and institutions.

First among them is Elliot Evans, Curator Emeritus of the Society of California Pioneers, to whom this volume and exhibition are dedicated. The pioneer scholar of California silversmiths, a notable collector of California silver, and a generous donor to the Oakland Museum, Dr. Evans has provided sound guidance, encouragement, and delightful, rollicking stories which have sustained us throughout this project. Edgar W. Morse, Professor of History at the California State University at Sonoma, editor of this volume, and consulting curator to *Silver in the Golden State*, has brought historical scholarship and his remarkable knowledge of California silverware to the project, resulting in a new understanding of California silversmiths. Accomplished articles by authors Donald L. Hardesty, Leslie Greene Bowman, Donald Chaput, Deborah Cooper, John W. Burke, and Gerald W. R. Ward have provided us welcome and telling insights. Additionally, Leslie Greene Bowman and Donald L. Hardesty have been especially generous with their time and expertise in the development of this project. Michael

Silver in the Golden State

Weller must be singled out for particular high praise; co-owner of *Argentum Antiques* in San Francisco, without his commitment to this project—through his loan of objects, his knowledge, his contagious enthusiasm, and his good offices generally—we could not have succeeded. The simply superb color photography is the work of Stephen Rahn. Ron Mortimore made many of the fine black-and-white photographs of silverwares which appear in this book.

The generosity of the many lenders to this project, and their willingness to share their splendid collections with us, have enabled us to present this exhibition and publish this book. To them go our profound gratitude and appreciation. Institutional support of the project has been exemplary. Special thanks are due to the Los Angeles County Museum of Natural History, the Los Angeles County Museum of Art, the California Historical Society, the M. H. DeYoung Memorial Museum, the Mackay School of Mines at the University of Nevada at Reno, the Bancroft Library, the Society of California Pioneers, and the Santa Barbara Historical Society.

Financial support for this project has been generously furnished by the Wells Fargo Foundation, the Women's Board and the History Guild of the Oakland Museum Association, the Homestake Mining Corporation, the American Decorative Arts Forum, Argentum Antiques, and several anonymous donors.

The staff of the Oakland Museum History Department has risen admirably to the challenge of *Silver in the Golden State*. Deborah Cooper has served superbly as project coordinator, seeing to her myriad of responsibilities with consummate skill, patience, tact, and tenacity. Thomas Curran, designer of this volume and its associate editor, has brought vision and high standards to both tasks, yielding a publication that is a pleasure both to read and to hold. Associate curators Inez Brooks-Myers and Mickey Karpas, conservator John Burke, preparator Carol Blair, technicians Kaoru Kitigawa and Robert Boudreaux, interpretive specialist Jacquelyn Goudeau, and secretary Seena Brown-Allison have each brought their special talents to the project. Last—and far from least—intern Britt Leggett and collection volunteers Joellen Lippett, Pat Carroll, Laura Hass, Joli Forth, Bob Gross, and Pat Monaco have contributed many and varied labors of love.

All that glitters is not gold!

L. Thomas Frye
Chief Curator of History
The Oakland Museum

Dedication:

Interview with Elliot Evans

DEBORAH COOPER

THE OAKLAND MUSEUM is honored to dedicate the book and exhibition *Silver in the Golden State* to Dr. Elliot Evans. A pioneer in the study of California silver, Dr. Evans has done much to bring the subject to the attention of scholars and collectors. His own prodigious talents as a collector led to the creation of two important institutional holdings of California silver: while Curator at the Society of California Pioneers from 1956 until his retirement in 1971, Dr. Evans built an outstanding collection of the work of the major nineteenth-century California silversmiths, and in 1981 the Oakland Museum acquired his personal collection of over 200 pieces of nineteenth- and twentieth-century California silver. As advisor to *Silver in the Golden State*, Dr. Evans continues his lifelong dedication to the study of California silver, material culture, and history.

In an interview with Deborah Cooper of the Museum, Dr. Evans related some of his early explorations into California silver.

How did you first become interested in silver?

When I was a kid, my favorite aunt would come out from Nevada where she taught and later owned a newspaper, the Yerington Times, *which she edited during the First War. Well, she would bring her silver travelling kit with her, which if I recall now, was something they got as prizes for saving cereal box tops. She had a big set of this stuff, dresser jars, brushes and whatnot, and I would help her clean it. After that, I always went for silver in a large way.*

I always wanted silver, but we didn't have much; in fact, we had very little. I think we had a set of violet pattern teaspoons which were my mother's wedding present, and a silver-rimmed cut glass jelly dish which used to appear at Thanksgiving and Christmas for cranberry sauce. When I got older a family friend, Mrs. Coffin, gave me a little pair of salt spoons. They were Gorham Grecian, and I now have 164 pieces of that pattern, possibly the only complete set in existence; it's very complete for dozens.

My wife Elizabeth's family was involved in mining and real estate in Virginia City. Her grandfather, Lewis Janin, had been mine superintendent at the Enrequita Mine in New

Dr. Elliot Evans

Thumler and co-workers at Shreve's

Almaden. They had quite a lot of silver, all premium ware from East Coast makers. We came into some of that when we were married, and so we just kept right on collecting it.

What intrigued you about *California* silver in particular?

Arthur Thumler was the one who put me on to this, years ago. I just sort of happened in off the street with a silver repair job for him, and we became friends. He was very angry because he said the WPA people had done a study of California silver, and the whole thing vanished. He was awfully distressed about it because he had helped out with his own data, he said it was all complete, typed up and bound. He was agitated about the subject, and put me on to it.

I did check with the National Archives about that report, and they don't seem to have a record of it. Did Thumler know the names of the people who were involved in producing the report?

No. They were just nice ladies. I finally ran one of them down at the deYoung Museum [in San Francisco]. She wouldn't talk about it; she would see me coming, and just make a beeline for the front office and close the door. I felt kind of sorry about it because it wasn't her fault, but I really think she was so embarrassed about the project because she had seen it dumped. You see, when the War came and those WPA projects were rapidly folding, the contents of their offices were dumped in the Bay, I understand, for landfill. Needless to say, nothing has turned up.

Thumler is one of the important silversmiths whose work is included in our study. You knew him well; what do you remember about him?

I have never known a man whose language was so profane. Every second word was a curse; the way he went on was incredible. But he was cheerful as anything, and you couldn't help liking him. He knew so much. And he felt so keenly that the deYoung Museum didn't even have a single piece of California silver on exhibit.

The deYoung did have a pair of bronze and wood empire-style candlesticks which Arthur admired exceedingly. He made a copy of them in silver, and I remember they were on display in his window. He drew inspiration from the Museum collections.

Thumler had worked at Shreve and Company before setting out on his own, hadn't he?

Yes. He worked there before the 1906 fire and afterwards. Shreve is one of the two or three longest-lived firms in California. At the very time eastern firms were consolidating branches from around the country, concentrating the manufacture of silver around Providence, Rhode Island, and nearby, Shreve decided to go into their own silver manufacturing business in this western outpost of civilization. And it paid.

There continue to be rumors that records from the early days of Shreve and Company exist somewhere, but haven't been found yet. Are the rumors true?

No, I don't think so. I had heard that Shreve was discarding things when they closed the factory in 1969. I went down there and pleaded with them to give their records to the Society [of California Pioneers]. *But the new owners weren't interested in history. Everything went to the dump.*

Without these records, how did you start piecing together the history of Shreve?

From city directories. This list shows Shreve employees from various periods, from the directories. I don't know how I got so many of them. It took hours and hours. For example, in 1888 there were six clerks, four salesmen, four "others," one correspondent, one porter, and one watchman in the Shreve shop. In the factory, there was a superintendent, one engraver, thirty jewelers, one watchmaker, and three silversmiths. Their names are all here. Fifty-three workers plus George C. Shreve and his three partners.

Were you able to find out anything about the silversmiths who worked for Shreve?

One of them was Frederick George Witt, and the "Bee" mark found on some Shreve pieces was his. I got Seymour B. Wyler's book of silver marks, and looked up all the marks that had a bee in them. I found out that the bee was much-used in Germany. Then I went all through the directory listings, year by year; I wanted a man who had a German name, whose family mark had a bee symbol in it, and who was listed as a silversmith or employee at Shreve. I found George Witt. He was from Germany. German trained. He belonged to German fraternal societies and so on. The dates all fit, and it seemed likely that the bee was his personal mark rather than having anything to do with Shreve.

Did other Shreve silversmiths put personal marks on their work?

There are two, maybe three pieces where they have shown up. But in 1894, shortly after George Shreve's death, the firm incorporated as Shreve and Company, and private marks were forbidden. Arthur Thumler told me that somebody put a private mark on some pieces without permission, and they made him take every mark off and put the Shreve mark back on. That was only sensible; after all, if it's an incorporated business you're working under their flag, not your own.

Not everyone agrees that the bee was a personal mark. Did Thumler mention it in particular?

No. I think you can't get around the fact that Shreve used it. And to what extent it was theirs legally, I don't know. I don't think it even really matters a whale of a lot except that it's got some caché. The puzzle is, where did they get the bee mark? That's the connection to Witt. I think that Witt used it when he was running the shop, as it were.

The silver objects themselves must have been sources of information. The collections you created are outstanding; how did you go about building them?

Did you ever hear of buying in shops? At the time I was collecting, I seemed to have more leisure and mobility to go around looking for things. There used to be a lot of secondhand stores on McAllister Street, beyond City Hall.

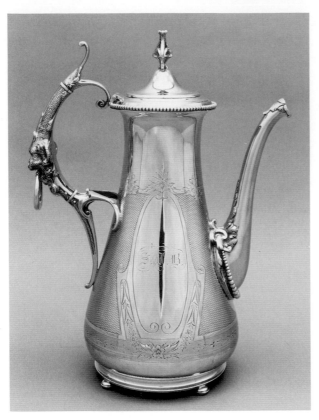

Plate with engravings of California symbols,
Schulz & Fisher, *c. 1880*

Blasdell coffee pot, collected
for the Society of California Pioneers

"Bee"-marked Shreve "Melon" tea set
collected for the Oakland Museum

You know Charles Fuller? Well, Charles was a wonderful help and a very good guy. Charles and I would exchange and accumulate. He'd find something, and he'd tell me, and I'd go and look. I think I gave him two or three pieces, or we traded some things. It just kept growing and growing.

Charles and I got that big plaque with the scene of Yosemite engraved on it (PLATE 11). *It was from somebody in the East, and we debated whether or not it was worth what they were asking. I think we beat them down a little bit, and then bought it, because we both realized it was a damn good piece, important to the history of silver. I think we both put up half or something like that. I was low at the time, but later I bought Charles out. He was always pleasant to deal with and knew that the chase would go on.*

It sounds like you weren't alone in your interest in California silver.

No. Grace Elliot was the one who first called attention to Vanderslice as the "Tiffany of California." She had an antique store in Los Angeles, and she would let me paw through the drawers of spoons. I would pick out what I wanted. We have gone on being friends.

Rev. Paul Evans (no relation) and I used to trade back and forth. He had a small parish out in Antioch, and we would meet at a shop called "Lila's" out in Lafayette to see if we could find anything, or if I had found something he wanted to look at.

And the chase would go on. Were there any big ones that got away?

There was a Koehler and Ritter tea set that came to Joe Baisa for repair. It was a beautiful set, and uncommon. The owners were thinking of selling, and instead of pouncing on it, I counted my pennies and decided I didn't have enough. Somehow or other, the idea of $300 for that tea set seemed a lot at the time. Knowing then what I know now, I should have gone into debt for it. That one got away.

Only one out of a great many pieces, though, that you did manage to save. Looking at California silver as a whole, is there something that sets it apart from other American silver?

I don't know, really. The craft standard would be about the same, so that wouldn't make any particular difference. And people looked to the East Coast for what was fashionable. Shreve always struggled to get the dealership for Tiffany silver; it lent prestige to the firm. California silver had to follow the eastern manufacturers in order to be popular.

If anything, subject matter is the difference. The images and symbols used were very literal —all those miners and grizzly bears, Yosemite, the mountains and the locomotives. Those images give a distinct flavor to California silver. There's nothing else quite like it.

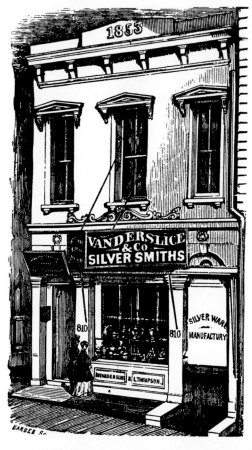

Silver in the Golden State

EDGAR W. MORSE

SILVERWARE WAS IMPORTANT to Californians long before there were California silver-smiths. The native Indians must have been awed and impressed by the wealth of silver the Padres brought out of Mexico to furnish the mission churches. Even in the poorest circumstances, the Padres understood that silver, in its purity and richness, reflected the brightness and majesty of God. On the ranchos too, silver was chosen over the equally utilitarian copper, tin, or iron, as a symbol of the wealth and power of the owner.

Some of the privately owned silver was an important indication of rank. The commander of the Presidio of Santa Barbara, José de la Guerra y Noriega, brought his silver desk set from Mexico (PLATE 1). General Vallejo's status is reflected in his elaborate silver *sahumador* (incense burner). Much privately owned silver is unmarked and may have been made locally, particularly small objects such as silver-mounted saddles, bits, and bridles, which were probably made in Alta California from earliest times. Typical of such local work might be the silver curtain stays of Señora Catalina Norena y Pico, the silver saddles and tack proudly displayed by Don Vincente Lugo (PLATE 2), and the silver-embroidered velvet jacket worn by Don Antonio de Coronel. By Spanish and, later, Mexican law, silversmithing was illegal except in such major centers as Mexico City, where taxes could be collected easily. But the law was widely evaded, and there never was an assay office established in Alta California.[1] Many silversmiths with Spanish names are recorded in the 1852 California census, but the work of only a few of them survives, notably that of Celestino Trujillo and José Guadalupe Avila in Monterey. The great majority of colonial work found in California is unmarked and cannot be documented. A rare exception is a mug recently discovered near San Luis Obispo that bears the engraved ranch brand of the horses of the Arroyo Grande Rancho in that vicinity.

The Padres also trained the Indians in metalworking at the missions. Surviving documented work is in copper, such as the baptismal bowl at Mission Santa Inez. The Indians at Mission San Fernando were also widely recognized as skilled coppersmiths.

Vallejo's Sahumador, Made in Mexico City, c. 1800.

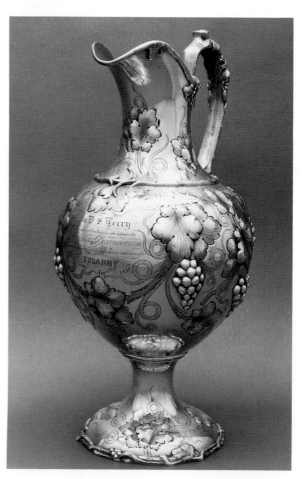

Ewer by George Ladd Presented to Justice Terry

Crudely raised copper vessels, as well as many copper scraps, have been excavated on the grounds of Mission Carmel. Anyone trained to make such copper pieces could have worked in silver as well, because the raising techniques are identical.[2]

Statehood and the Gold Rush brought many silversmiths to California, but only a very few of them worked at their trade, as evidenced by the rare coin-silver spoons, indistinguishable from work done in the East, marked by various San Francisco silversmiths. Silverware may also have been made outside San Francisco as early as the 1850s, because spoons exist marked, "EAVES & NYE Sac[ramento] 134 J. St." The English-born Eaves had worked in Cincinnati, Lexington, and St. Louis before arriving in California in 1849. His spoons are in a style peculiar to the Midwest. Another early jeweler, William A. Woodruff, began advertising "Woodruff's Jewelry Shop" by August, 1850, and is said to have made the gold chain given to Daniel Webster by George W. Egleston. More likely, Woodruff actually had it made by one Charles Hendrick, who was listed as a goldsmith in the 1850 directory and as a silversmith in the 1860 census. When Woodruff died in 1850, his business was absorbed by William, Hamlet, and Paulaski Jacks, his brothers-in-law and partners. Their firm was first styled "Jacks & Woodruff," later "Jacks & Brothers." Hamlet had arrived from New York in April 1849, and he and his brother William were listed in the 1852 census as jewelers. By January 1851 they were well enough established to be awarded the contract for a thirty-ounce solid gold goblet that was awarded to the Collector of Customs for Upper California. Jacks & Woodruff were considered to be "quite sound" in June 1850 by one of R. G. Dun's informants. By February 1854 they were no longer listed in the directories.[3]

John W. Tucker, one of San Francisco's earliest and most prominent jewelers, claimed to have made in 1854 a large gold service for Mayor Cornelius K. Garrison. It consisted of fifty pieces, and cost more than $50,000 to manufacture. Tucker was a flamboyant advertiser, a successful retail jeweler, and a mining and real estate speculator, but it is likely that he commissioned this set from an East Coast manufacturer. His own training had been as a carpenter, and there is no evidence that he employed silversmiths, let alone ones capable of an undertaking of this scale or quality. Although much silverware exists bearing his name, it all appears to have been made by manufacturing silversmiths such as Gorham or Vanderslice.[4]

George Ladd was another retail jeweler who appears to have put his name on unmarked Eastern-made silver. Ladd was born in England and moved to New York when he was seventeen. Initially he worked for his brother, William F. Ladd, who was to have a prominent career with the New York City firm of Gelston, Ladd & Co. through the 1880s. He then set up business for himself for a few years. In common with many other young jewelers, Ladd sought new opportunities in California. He arrived by steamer in 1852 and worked in the trade till 1856, when he set up his own shop with his brother's assistance. Within the year, Ladd had provided a large silver ewer that was presented to Judge David S. Terry by the "ladies of San Francisco." Though it bears his stamp, there is no independent evidence that he had the skill to fabricate this elaborately formed piece himself, or employed anyone else who did. Presumably the ewer was supplied by his brother from New York, because Ladd continued to advertise in 1858: "just received a fine assortment of . . . Watches, Jewelry, and Silverware." But in January, 1861, Ladd announced that he was "Selling Off" everything and "going East in April. . . ." He later became an expert diamond salesman for Tiffany in New York.[5]

Other prominent San Francisco jewelers put their names on unmarked silver of other makers or added their retailer's stamp to pieces marked by Gorham or Vanderslice. One such was the firm of Barrett & Sherwood, who began business in 1849. They were pioneers in making gold-quartz jewelry, which they exhibited at the Crystal Palace Exhibition in New York in 1853. There is no evidence that they employed silversmiths. Another firm, Braverman & Levy, began about 1852 as retailers of jewelry, watches, and silverware. Most of the silver bearing their mark was made and marked by Gorham.[6]

William Lawler was the first arrival from the East who can plausibly be called a working silversmith, capable of making his own wares. He was born in Ireland in 1809, and worked during the 1840s as a watchmaker and silversmith first in St. Louis, and then in New Orleans. Married and with three small children and a servant in 1850, he seems to have prospered until the lure of California brought him to San Francisco by February of 1854. He may have left his family in New Orleans, for he continued to be listed in that directory until 1855. He appears in the San Francisco census for the first time in 1870 living with one son, Frederick. His wife and other children have left no trace. Lawler had a fitful career in San Francisco. At the first San Francisco Mechanics' Industrial Exhibition in 1857 he exhibited gold, silver, and enameled work, for which he received a diploma. He sustained a business in his own name until about 1860, when his shop was no longer listed in the directories. For some years he was a "gold and silver worker" for others, a boiler maker at the Union Foundry, and a gold-pen maker. In 1863 he opened his own shop again, but failed within the year. Finally, in 1867, he established another shop at 653 Clay Street, and began to advertise as a "pioneer silversmith and jeweler."

By 1870 Lawler was operating a respectable small shop. His only employee was his son Frederick, but together they produced some $20,000 of gold and silver work. They used hand-power and had no machinery of any consequence except a forge. But his success was short-lived. Frederick disappeared from the directories after 1876, the shop had ceased to exist by the 1880 census, and Lawler himself died in 1881. His use of 300 ounces of gold and 5,500 ounces of silver in 1870 suggests that examples of his work ought to be more plentiful than they are.[7] He made plain fiddle-pattern flatware of various sorts, which is the only silver known by him with the exception of a small mug at the Oakland Museum. After a somewhat promising career in the Midwest and the South, Lawler seems to have gambled on making his fortune as a silversmith in the Golden State. He lost the gamble. The first man to make a success of it was Vanderslice.

In 1858, William Keyser Vanderslice established the longest lived of the nineteenth-century San Francisco manufacturing firms, and made a fortune as a result. Born in 1823 in Philadelphia, Vanderslice was educated there and learned the trades of a closeplater, electroplater, silversmith, and jeweler. In 1847 he married Kate Sherman, a member of a branch of the prominent Boston family. Although his name had appeared in 1840 as a silverplater (closeplate), he probably worked for his brother-in-law Samuel Hopper, a jeweler, until he began a partnership with one M'Cully, a watchmaker, in 1857. On December 4, 1857, he took a clipper ship from San Francisco with his family, his tools, and $206 of borrowed capital.[8]

By 1857, the California gold rush was over, but San Francisco had become sufficiently wealthy and sophisticated to support jewelers, silversmiths, and other purveyors of luxury goods. Vanderslice's decision to go West surely must have been based on reports from his brother-in-law, Charles H. Sherman, who had become established in San Francisco as a dealer in carpets and upholstery as early as 1852. For the first two years after his arrival, Vanderslice located his shop in Sherman's building at 134 Washington Street. By September, 1860, he had expanded his operation and moved to 728 Montgomery, increasing his capital by forming a partnership with Sherman, which lasted until 1865. For a short time he also employed another brother-in-law, Leander S. Sherman, who would become a socially-prominent San Francisco merchant.

Vanderslice seems to have clearly understood how to establish his reputation both by exhibitions and by advertising. In September, 1858, four months after his arrival, he exhibited a pitcher at the Industrial Exhibition of the Mechanics' Institute. The judges reported that it was "... very prettily designed ...," and deserved a diploma. One year later he won a "First" premium for his exhibit at the California State Agricultural Fair. The award committee was obviously impressed when they reported:

> We have carefully examined the articles by W. K. Van Derslice (*sic*), as his own manufacture, and we find the workmanship fully equal to that of imported articles. The large pitcher and two cups to match we consider worthy of notice, from the novelty of design, and the beauty and excellence of its execution. ... We recommend for the above the first premium.

Vanderslice did not compete again, but by 1872 he had begun to receive large annual contracts to supply the Fairs' silver awards. He advertised as far away as Victoria, British Columbia, where in 1860 he was prepared to sell all kinds of solid silver articles to "The Trade ... ," who no longer needed to "... send to New York for articles in this line. ..." He proposed to supply everything from "threaded or plain" flatware to tureens and waiters. The firm's growth was rapid. An article published in January of 1865 describes his factory and a number of the machines that he used to roll silver into sheets, spin holloware, and form flatware handles by roller dies. Starting with one employee in 1858, a decade later he employed twenty men and used about 1,500 ounces of silver a month. The addition of a new partner in 1868, Lucius Thompson, seems to have provided some of the money needed to finance this progress.[9]

Thompson, a jeweler and watchmaker, had been a partner of George C. Shreve's for the preceding five years. He brought new skills as well as new money. He probably had about three times Vanderslice's capital, considerable experience as a retail jeweler, and was a capable business man. Vanderslice, in turn, was "the practical man in the shop" in addition to being socially well-established through his connection to the Sherman family. Although profitable, this combination of assets would lead to major changes in the business. At about this time, Vanderslice applied for his first patent for the flatware pattern now known as "Gargoyle." This was the first indication of a larger plan for expansion, using Thompson's new money to pay for dies and machines to produce new flatware patterns.

By late 1869, the R. G. Dun reporter was quite approving of the partnership. "Both v'y steady indu[strious] men, g'd hab[it]s . . . In g'd cr[edit] but their bus. reqs li[ttle] is (*sic*) any cr[edit]." During the year preceding the summer of 1870 they produced some $70,000 of a variety of solid silver products using some $21,000 in silver and paying $13,000 in wages to twenty workers. Working capital was now about $35,000 invested in tools, dies, and in an eight-horsepower steam engine that powered some twenty machines including cutters, rollers, and lathes. By 1871 Vanderslice was recognized as doing ". . . the lead[ing] bus[iness] in their line on The Coast. Are E[asily] w[orth] in stock. Mach[iner]y & cash fully $50,000." The next step in their expansion was a move to 136 Sutter Street in the heart of the fashionable retail district. Along with this move, in 1872, came an accelerated shift from manufacturing to retailing. Within two years Thompson was going East to buy jewelry, clocks, and watches, although they did not change their apparent practice of buying the major share of their jewelry from Levison Brothers/California Jewelry Company, who were also Vanderslice's landlord. Advertising of the expanded retail venture began on January 1, 1875. "Plated Ware! The largest Assortment in California . . . , Jurgensen Watches . . . , Peregaux, the Celebrated Watch . . . , Jewelry of all kinds at manufacturers' price. . . ." The move, both in location and type of business, was entirely successful. R. G. Dun's reporter observed in 1875: "Nearly 3 yrs ago when they moved to present location they showed a net surplus of $58,000 in their bus[iness] taking everything at close figures. The first year after moving their Sales nearly doubled & they have done a steadily increasing bus[iness] ever since and are now considered to be wor[th] $75,000 at a safe estimate. . . ." They continued to patent new flatware patterns, the second being "Comstock" in 1874, but the Dun records indicate that they were buying extensively from Gorham and Whiting at the same time. Throughout the 1870s the factory's own production remained fairly constant. But financial problems outside the firm's control would soon have a major adverse effect on the manufacturing end of the business.[10]

The depression of 1873 reached California in 1875. On August 26, the Bank of California suspended all payments and its president, William Ralston, drowned that same afternoon. Virginia City burned October 26th., and the plants of the Ophir and Consolidated Virginia mines were destroyed. Mining stocks on the San Francisco stock exchange fluctuated wildly, but it soon became evident that production on the Comstock was about to collapse. From a peak of $33.7 million in 1876, it fell to $7 million in 1879 and a dismal $1.9 million in the first half of 1880. A financial disaster of this magnitude would inevitably affect the merchants who had come to rely on the purchases of luxury goods by those who had made fortunes in mining and other speculations.[11]

Vanderslice's position deteriorated rapidly. In 1878 he bought out his partner Thompson for "Cash and notes." By the end of the year he began advertising an "Immense reduction in silverware. Solid Sterling Silver Spoons and Forks of our own manufacture, at $1.40 per oz, it being much lower than ever they have been sold." At the same time, Shreve was offering sterling spoons and forks at $1.50 per ounce, which was a 7% higher price. The next year's factory production was only $20,000 compared to the $70,000 a decade before. The work force declined sharply from twenty men to five. A few months later, in July and August of 1880, Vanderslice put up "Our Entire Stock" of goods at a public auction at their shop. These sales were part of a program that Vanderslice's financial officer, W. R. Drake, had started in 1878, when Thompson left the firm. According to Drake: "I was a sort of financial manager and negotiator for Vanderslice and Company. They were in debt about $130,000 and I went in to straighten it out. An Auction sale and a great many other things were done to extricate them from that difficulty." The worst of this crisis seems to have passed by the end of the year. Their advertisement in January of 1881 was entirely optimistic, and sought to correct the impression that Vanderslice was going out of business. They advertised their own flatware at $1.55 per ounce, an 11% increase, as well as the arrival of luxury goods from the East. Perhaps the association of a new partner at this time, Kenneth Melrose, financially assisted the firm.[12]

The days of Vanderslice's large-scale silver manufacturing were over. He continued to advertise as a silversmith, but his work force during the 1880s and 1890s averaged about four men, and dropped to one at the turn of the century. By 1892 several of his men had moved to Shreve & Company. Vanderslice's holiday ads, instead of emphasizing silver, talk about "gold quartz jewellery (sic), diamonds, watches, and fine jewelry," along with opera glasses, rhinestone jewelry, and "All the *Latest Novelties*...." As "the practical man in the shop," Vanderslice's role was clearly changing. Then, on May 8, 1893, he fell down a flight of steps, suffered a compound fracture of his left leg, and had the leg amputated. He seems to have had a stroke early in December, 1895, and he died March 2, 1899.[13] His business was to survive him for a time, under various owners, mostly as a fancy goods store. By that time, the firm had lost touch with its heritage as silversmiths. In 1896, for example, a Reno jeweler asked Vanderslice & Co. to identify a spoon that bore their mark. They could not recall which "Eastern" firm had made what was in fact Vanderslice's own "Waterlily" pattern from the early 1860s. But as a retail jeweler, the firm was undoubtedly a success.[14]

April 18, 1906 changed everything. After the insurance claims had been sorted out, they were found to be only $400 coverage on machinery. Twenty-six years had seen a $35,000 investment in tools and machinery reduced to a claim of $218.12, which was paid in full. The end of the firm was announced on November 28, 1908: "SHREVE & CO. WILL ABSORB OLD FIRM. W. K. Vanderslice Company will be taken Over Immediately after the holidays." In January its stock was sold at reduced prices and the Vanderslice store closed. For a few years after the merger, Shreve included the Vanderslice name on its invoices.[15]

Vanderslice's production was large and varied during the "good" years. Between 1858 and 1885, he introduced at least sixteen flatware patterns, most of which were full lines. He made objects ranging from napkin rings to gigantic epergnes (PLATE 6), of quality comparable to that of the big Eastern firms. No holloware marked 134 Washington Street (1858–1860) is recorded; pieces marked 728 Montgomery Street (1860–1863) are very scarce. Enough objects marked 810 Montgomery Street (1863–1871) exist to permit some generalizations. He followed the prevailing "renaissance revival" style in most large pieces, characterized by the use of animal head castings, together with a "broken" Greek key ornamental border. A pitcher presented to A. J. Ralston in 1870 is a typical example, with a grotesque-head handle, beading, and very elaborate and accomplished

W. K. Vanderslice, *Pepper Mill*

W. K. Vanderslice, *Cruet Set*

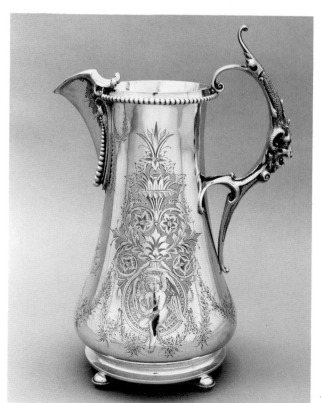

W. K. Vanderslice, *"Ralston" Pitcher, 1870*

engraving. Much of this work is frankly copied directly from contemporary work by Gorham, Ball Black, and Tiffany. In this same period, he made some rather severely plain pieces with octagonal sides from flat sheet stock in a kind of Queen Anne-revival style. But Vanderslice was also a stylistic innovator. The illustrated cruet set and epergne are in no style but his own, with their gear-tipped feet indicating a frank Victorian celebration of the machine. The pepper mill is another unique design: poised on its ivory tipped tripod, it almost anticipates a rocket about to be launched into space. (The steel works for this object were made and marked by Will & Finck, the noted San Francisco cutlers, who had a sideline of making gamblers' cheating equipment.)

Pieces marked 136 Sutter Street (1872–1906) are relatively less common and considerably less interesting in style. They are often made of thin sheet silver, and seem to have been made to meet a price instead of an esthetic ideal, rather like the work of their low-end competitors, Schulz & Fischer. Typical of their later work is a very fine silver mounted cut glass box, where the chief value surely was in the glass. Vanderslice supplied, in 1898, a "cut glass bowl mounted with 500 Klondike nuggets," for the Alaska Commercial Company. It was one of the firm's last productions, and it reveals something of the change in the nature of the business. From innovative silversmiths, Vanderslice & Company had become prestigious jewelers who soldered someone else's gold nuggets together to mount someone else's cut glass.[16]

Vanderslice had several competitors of considerable skill who are almost forgotten today. Their obscurity is probably a result of the short-lived nature of some firms, and because other firms catered to somewhat "marginal" customers. The major firms were Friedrich R. Reichel, Koehler & Ritter, and Schulz & Fischer. Unlike Vanderslice, none of these people were in the social register, and none advertised in the San Francisco *Blue Book*. Curiously, they were all Germans, and their lives and products are as interesting as those of the more prominent silversmiths.

Reichel was born in Germany about 1824, and had established himself in San Francisco as a jeweler by October, 1856, working in the trade. In late 1861, he first advertised himself as a manufacturer of silverware as well as jewelry. His business grew swiftly, and two years later he hired William Schulz and Emil Fischer, both silversmiths, and Charles Ritter, a bookkeeper. The following year he added Gotthard Koehler, a jeweler, who acted as his foreman. Reichel's business seems to have prospered until his death, unmarried, at age 43 in 1867. He had at least six employees in 1863 and eight in 1867. He left, at his death, the substantial estate of $28,634.78. With the Reichel firm dissolved, Koehler & Ritter went on to advertise themselves as "Successors to F. R. Reichel," and Schulz & Fischer formed a separate partnership. Little Reichel holloware seems to have survived. The water pitcher in PLATE 3 bears his mark and was probably made in his shop. Although the renaissance revival decoration is derivative, the details are unique to Reichel. Seven of his flatware patterns have been identified, all struck using double dies, which indicates that he had a well-equipped shop and access to a talented die sinker. The designs are not distinctive, but there is no doubt he did produce them himself.[17]

Gotthard Koehler and Charles A. Ritter, successors to Reichel, became one of the principal nineteenth-century San Francisco manufacturing firms. Koehler was born in

Saxony in 1823, lived in Oregon in the early 1860s, and finally settled in San Francisco in 1864, where he was naturalized in 1867. Ritter was born in Hesse Darmstadt about 1840, settled in San Francisco in 1864, and was naturalized in that same year. When Reichel died, they bought his stock and machinery for about $10,000 and continued to operate at the same address. They got $8,000 of the purchase price as a loan from the manufacturing jewelers, Levison Brothers. By the mid 1870s they employed from twelve to fifteen workers, had invested capital of about $30,000 and an elaborately-equipped factory. Powered by a ten-horsepower steam engine, they had two lathes for spinning, a flatting mill, spoon roller mill, cutters, presses, a steam drop press, a furnace, and a forge. In the year from mid-1869 to mid-1870 they used about 9,500 ounces of silver and 400 ounces of gold to produce $45,000 worth of silverware and jewelry. Koehler acted as the foreman and favorably impressed R. G. Dun's informant who reported that: "... he is steady, indust[rious] saving & bears an excellent char[acter] in every respect—has very limited [financial] capacity & pays no attention to the management of the bus[iness]." Ritter, in turn, was the wealthier partner and acted as: "... book keeper, salesman & manager—is steady economl. & attentive." The reporter, however, had little that was good to say about Ritter's character: "... possesses fair business capacity, but is rather quick tempered—keen & shrewd in looking after his own interests—unscrupulous & unreliable—will take advantage whenever opportunity offers." Worse was to come. "There are rumours afoot respecting his management of affairs after 'Reichels' (*sic*) death, which are damaging to his char[acter]."[18]

These rumors notwithstanding, Koehler & Ritter continued to prosper during the early 1870s. They needed little bank credit, and by 1875 they were doing a retail business of $80,000 to $100,000 a year and had a net worth of probably $25,000. Rent on their premises was a mere $360 per year, less than one-half of one per cent of their gross sales. Yet their quite unfashionable location among many manufacturing jewelers was incompatible with the retail market they sought to serve. In late April, 1875, they completed a massive punch bowl and tray for presentation to William F. Babcock, former president of the San Francisco Spring Valley Water Company. With a bowl sixteen inches both in diameter and height, a portrait medallion of Babcock, a pedestal and tray to match, the "... two pieces are the largest ever made on this coast, the combined weight being 460 ounces, and the cost about $3000." In their first exhibition at the Mechanics' Fair in the summer of 1875, they won a silver medal for "Chasteness of design, original modelling, and beauty of workmanship." In 1876 the report was even more favorable: "Made a remarkable fine showing of the skill and taste attained in this city in the manufacture of silverware. Their display consisted of almost everything required for use or ornament." With such recognition, it must have seemed a wise decision to move their combined retail store and workshop to a more fashionable situation. Their new premises at 26 Post Street put them in the "right" location, but it also increased their rent more than eighteenfold to $6,600 per year. Sales seem to have increased little if at all as a result, however. Their business was now about $108,000 per year, but with the effective rental for these sales increased thirteen times, their profits could only have declined.[19]

For a while they continued to receive important commissions and reviews. In 1878 they exhibited at the Mechanics' Fair an "EXQUISITE AND UNIQUE EPERGNE,"

Koehler & Ritter, *Butter Tub*, 1885

Koehler & Ritter, *Centerpiece*

Koehler & Ritter, *"Masonic" Tea Set*, 1885

that had been made for "THE WIFE OF ONE OF OUR MONEY KINGS." It must have been an extraordinary piece: "While there in no timidity in the handling, there is evinced that good taste which always knows when to leave off at the right time... Their popular store on Post Street is the resort of the elite of the city, and their name is a household word among the people of San Francisco." Shortly after their move to Post Street, Koehler & Ritter had also planned to greatly enlarge their manufacturing capacity, "but this is looked upon as rather unwise & not likely to benefit their bu[siness]". The R. G. Dun informant gave good advice, for in March of 1879 their stock was attached for $70,000 and their store closed, all on the order of their original financial supporter, Levison Brothers. An agreement was reached to continue operations, and in the year ending in mid-summer, 1880, they had at least thirteen employees, $40,000 invested in manufacturing, and used $25,000 in material to produce goods worth $50,000. But the agreement collapsed, and late in 1880, J. T. Bonestell of Levison Brothers announced that he was holding an "Absolute Closing-Out Sale of the former Stock of the late firm of Koehler & Ritter." Koehler & Ritter made an attempt to salvage a business from the debacle, but it was not to last. Koehler disappeared from the April, 1885, directories, and the firm was now styled "Charles A. Ritter, formerly Koehler & Ritter." Ritter seems to have retired by early 1891, and died some years later. Koehler literally disappeared: in his seventy-second year (1895) he had "left his home for a walk and it is feared that he wandered down among the wharves and fell into the bay." This dismal business history should not obscure the fact that the firm produced a large quantity of silverware, mostly flatware, as well as occasional pieces of specially commissioned holloware. They introduced at least six new flatware patterns, and continued to make three of Reichel's. Their holloware is relatively scarce, is of moderate- to heavy-gauge, and is usually very well-made. At times, their designs are quite distinctive, as in the illustrated butter-tub, but there is also evidence that they cast details directly from pieces by Gorham. One of the few large pieces extant is the large centerpiece bowl with a figural stem and birds perched on the rim, which is a close copy of a Gorham design. Many interesting pieces are described in newspapers of the day, but are now lost.[20]

The other successor firm to Reichel was founded by two more of his former employees, William Schulz and Emil A. Fischer. Schulz was born in Prussia, arrived in San Francisco in October, 1863, and was first employed by Reichel as a jeweler, and later as a silversmith. Fischer was from Saxony, was living in Ohio in 1857, and arrived with his family in San Francisco in October, 1863. He, too, began working for Reichel as a silversmith, and unlike Schulz, was naturalized. Within a year of Reichel's death, they had organized the firm of "Fischer & Schulz." It was a marginal operation at best, since the other ex-employees, Koehler & Ritter, had managed to borrow the money to buy Reichel's tools and equipment. Within a month of opening, Schulz & Fischer found their own financial backing by adding a third partner, Christof F. Mohrig. Mohrig, a Prussian, arrived in San Francisco in 1856, and began his career as a jeweler and watchmaker. He seems to have left San Francisco for some years, and returned from parts unknown with enough money to join Schulz & Fischer's firm and to establish his own factory for making gold and silver watch chains, which proved very profitable. One year later, in a report that R. G. Dun made to Gorham, Schulz & Fischer (but not Mohrig) were described as doing "... a sm[all] bus[iness] have but mod[erate] means—nice men: v[er]y indus[trious] hard-working & attentive. Hon[est] and hon'ble in their dealg's. In gd cr[edit] for mod amts." From a small starting capital of $5,000 invested entirely in tools and machines, they made substantial progress in the 1869–1870 period. According to the census of that year, they now had $14,000 in tools and inventory, and rented forty horsepower of steam energy to operate a roller mill, a spoon mill, a cutting press, and a lathe. They employed six workers and used 7,000 ounces of silver to produce $22,400 worth of flatware and small holloware.[21]

Their first important recorded commission was for the finishing in 1869 of two gold spikes used in the ceremonial completion of the Transcontinental Railroad (PLATE 5). This was actually a sub-contract from David Hewes, who had had the spikes themselves made by the Vail blacksmith shop. By February, 1873, Schulz & Fischer had done well enough to buy out Mohrig's interest for $5,400. In addition to their own manufactures, they continued to buy from Gorham, and at least one extant tea set bears their mark but appears to be identical to a design by Gorham, except for different applied die-rolled ornamental bands. The next step in their progress came late in the Spring of 1875, when they changed from the "coin" to the "sterling" standard, and opened a sales office at 513 Market Street. This summer also saw their first entry into competition at the Tenth Mechanics' Fair exhibition. They won a diploma for work that was adjudged to be ". . . superior in engraving, original designs, and fine work." Their exhibit and factory were described (and praised) in a nearly full column article in the *Mining and Scientific Press*, which especially noted their "satin finish." That characteristic finish is a common feature of their holloware. In addition to its esthetic appeal, it eliminated the labor required to produce a highly polished surface. The writer goes on to praise their use of modern labor-saving machinery and their skill in the ancient handcraft tradition. The praise was mainly directed toward the vast increase in labor efficiency that had taken place over the past twenty-five years. Instead of a rate of four spoons a day, a workman at Schulz & Fischer's factory produced thirty-six, an improvement of 900%. No silver-plated wares marked by them have been recorded, but they certainly sold it, since by February, 1876, R. G. Dun began furnishing reports to the giant silver plate manufacturer, Reed & Barton. In common with other San Francisco firms, Schulz & Fischer had begun to feel the competition both from electroplated and solid-silver goods from the Eastern manufacturers.[22]

For the next several years, though they continued to exhibit at the Fairs, they seemed to give most of their attention to flatware production. A leaflet of theirs that was probably published in 1878 illustrated fourteen patterns. For some of them, at least, they offered nineteen spoons, fourteen forks, twelve knives, six ladles, five tongs, and many miscellaneous pieces, representing a very large investment in die sinking. The work is often very detailed and is always skillfully executed. Most patterns were derivative, but some, such as "Faralone" and "Cleopatra," are quite distinctive. The same leaflet offers to supply a prodigious array of holloware—at least as wide in range as that of the major Eastern makers. Most likely, in common with their contemporaries, some of what they sold were the finished goods of other makers marked with their own stamp, and some were their own. Schulz & Fischer, as R. G. Dun's informant pointed out, catered to a less

Schulz & Fischer, *"Japanese-style" Bowl, c. 1880*

Detail, the Mary Stanton Barron service, Schulz & Fischer, *c. 1875*

Tea set closely imitating a Gorham design, Schulz & Fischer, *1885*

well-to-do clientele than their San Francisco competitors. Their work is often of lighter weight, but they added considerable strength to some of their holloware by the skillful use of ribbing, fluting, and other clever design and construction techniques. Since silver was commonly priced by the ounce, these devices enabled Schulz & Fischer to undercut the opposition for objects that served a similar function.[23]

Some of these techniques can be seen in the illustrated holy water bucket (PLATE 7), or "stoup," which was evidently made for the immigrant Catholic population, where its baroque style would have been immediately familiar. The basic form is made from thin sheet stock, constructed in several pieces by spinning. The decoration and ribbing were hand done by embossing. Not only did these features add elegance, they also added strength to compensate for the thin metal. The bucket has a spun liner that insures that it remains water-tight even today. Though this must have been a special order, they used a rather incongruous renaissance revival design for the handle. Some of their holloware uses elements from their flatware pattern dies, such as their cruet set using handle dies from the "Medallion" pattern. The footed bowl with applied fruit shows other aspects of their skill. The basic forms are spun and soldered together. The hammered surface, however, was produced by embossing, in which the bowl was filled with pitch and each faceted surface formed by hand using a special punch and a hammer. This is one of the very few San Francisco examples of the "Japanese" style that became popular in the East around 1880. Sometimes rather simple forms would be decorated with spectacular engraving, of which the finest surviving example is a plate engraved with scenes of Yosemite and other California natural wonders.

By 1882, Schulz & Fischer employed some twenty-two workers and were considered to be the ". . . heaviest manufacturing house . . ." among San Francisco silversmiths. In 1885 they had invested $50,000 in their business, employed sixteen men and three apprentices, and paid $13,000 in wages. Throughout the 1880s they continued to invest money in die work to produce one new flatware pattern each year. They seem to have been very competent and innovative silversmiths. If they had limitations, these probably would have been a result of their lack of capital and business experience. In 1883 they took in another financial partner, Samuel McCartney, who was a wholesale liquor dealer before, during, and after his partnership with them. Flatware patterns made after 1888 are marked "Schulz & McCartney," evidently indicating that Fischer had ceased participation in the firm. Fischer died January 4, 1890, and the need to liquidate the assets of the partnership to settle his estate probably left Schulz with too little capital to continue an otherwise innovative and successful business on the same scale. He continued on for a few years, and disappeared from the directories in 1900. The firm's decline was probably also a result of factors entirely outside of their personal control.[24]

Competition from Eastern manufacturers was becoming troublesome. For many years, the San Francisco industry had the advantage of having a vast local supply of bullion and a low transportation cost by sea or rail to their markets on the Pacific Slope. Freight rates on the Transcontinental Railroad were sufficiently high to discourage competition. By the late 1870s, however, this was beginning to change. Mass-produced silverplated wares were aggressively merchandised everywhere in direct competition to solid silver. Reed and Barton, for example, had obtained the order for the silverplated ware for the Palace Hotel in 1875, and by 1885 they had an active representative in San Francisco. Gorham had established a branch office and salesroom in San Francisco by 1879. Combined with this direct sales competition, dramatic reductions in freight charges were eroding one of the main advantages the San Francisco manufacturers had over the Eastern companies. From $6.50 per hundredweight in 1870, the rates dropped to $2.50 in late 1876, and again to $1.50 in 1885. Patriotism was becoming the only reason to patronize California manufacturers at the same time that the telegraph and the railroad were fast ending California's "colonial" status and helping to create a national consciousness as well as a national taste. Thus died the last outpost of nineteenth-century regional silver production anywhere in the country. The surviving firm, Shreve, began the new century with a different kind of business and a different kind of silversmithing, more suited to the modern age.[25]

George Choates Shreve and his half-brother Samuel began a jewelry business in San Francisco shortly after they arrived in 1852. George was born in Salem, Massachusetts, and worked for a time in another half-brother's (Benjamin's) jewelry store. Benjamin became the "Shreve" in the prominent Boston firm of Shreve, Crump and Low. George and Samuel shipped as sailors and then worked as wholesale jewelers in New York City until they left for the West. Their jewelry shop was well established by mid-1855, and offered a wide range of European fancy goods as well as "Every description of California Jewelry manufactured to order." Their claim to be manufacturing jewelers may have been exaggerated, for they are shown as employing only two watch and clock repairers in the 1870 census. They also were importing jewelry from Benjamin, at that time a partner in Jones, Ball & Co. of Boston. In late 1857 or early 1858, Samuel was drowned at sea, and it became necessary for George to replace him. Beginning in 1859, one Lucius Thompson is listed as Shreve's partner; he had been an employee of the brothers probably since 1856. Already George had done well. By 1860 he had property valued at $25,000, and seven years later the partnership had assets of $100,000. In 1867 they advertised that they made a specialty of selling Gorham's solid silver and electroplated wares. The investigator for R. G. Dun must have been very impressed when he reported in 1873 that: "They carry from 80 to 100,000$ of stock. Keep the finest goods in their line, have the cream of the City Trade and are making $."[26]

Until the 1880s Shreve concentrated his attention on retailing. But early in 1881 he increased his capital by adding a partner, Albert J. Lewis. Shreve began advertising in 1881 his: "New Departure. To meet a constant and growing demand for a finer class of jewelry than is usually imported to this city, we have established a new factory, complete in all the details of improved machinery, for the manufacture of the best goods that can be made." Their work force had reached a total of seventy by late 1882, and by February, 1883, they had made "Extensive Improvements in our Factory, and added largely to its Facilities. . . ." Shreve's silversmithing finally began in 1883. They began to advertise that they were "Manufacturers of Fine Jewelry and Artistic Silverware" in September. This same advertisement illustrated for the first time their well-known "Bee" trademark for their silver. One of their very first commissions must have been the close-order drill trophies that were awarded at the Grand Encampment of the Knights Templar in August of 1883, one of which is illustrated (PLATE 8). Major local and na-

tional publicity such as this was a fitting start for an eighty-five-year career as California's best known firm of silversmiths.[27]

But disaster swiftly followed. On October 8, 1885, their factory was destroyed by fire.
They were adequately insured, and by April of the next year they had moved the factory
and were again advertising as manufacturing jewelers and silversmiths, in the meantime
adding a number of former workmen from Vanderslice and Schulz & Fischer. Within
less than ten years, Shreve had become the effective successor of all the San Francisco
silversmiths. George Shreve died a very rich man in 1893. His fortune of some $250,000
stood as a monument to a shrewd merchant who knew how to benefit from the desire of
the City's nouveaux riches to display jewelry and silverware as an eloquent testimony
to their success in the Gold Rush, the Silver Rush, and the rush to settle California.[28]

Shreve & Company was incorporated in January, 1894, under the financial control of
Lewis, who had long-since owned a majority of the stock. During this period, the factory
seems to have concentrated on special orders rather than routine production. According
to the reminiscences of one of their employees who started work there in 1898 as a
"cash Boy":

> In my spare time I would work on the bench, shaping silver or soldering. The factory turned
> out a variety of special work in silver and gold. We would buy our gold and silver directly
> from the Mint, then we'd alloy it and roll it into sheets for flatware. Our people could make
> anything imaginable in hand-crafted precious metals or custom-designed jewelry. When
> President Theodore Roosevelt visited San Francisco, the citizens presented him with a
> 10-inch high golden bear, cast from solid gold from Shreve's.

A number of these spectacular orders are recorded, including the gold cup worth $5,000
presented to Solano County at the Mid-Winter Fair of 1897, and the laurel wreath in
silver and gold awarded in 1898 to the leader of the San Francisco Symphony, Fritz
Scheel. Shreve continued to prosper into the new century. Long since established as the
premier quality jeweler of the West Coast, their production at this time was sold only
through their own retail shop. By 1906 the firm was worth more than $1,120,000. With
the possible exception of souvenir spoons, it is clear that the factory was not producing
any proprietary flatware patterns, because their first catalog, entitled a "Suggestion
Book" and issued in 1904, illustrates only one pattern, "Iris," by Durgin.[29]

The 1906 Earthquake and Fire was a physical, but not a financial, disaster for Shreve.
Their uninsured loss of less than 5% of their assets was minor. Always alert to the value
of publicity, Shreve surely made it easy for newspapers to print articles that portrayed
their efforts to recover. For example, in the April, 1907, issue of *Sunset*, Shreve is quoted
as saying:

> When the fires were put out there was work to do. The first in importance was to construct
> a stockade, eight feet high, around the vault, and to have a sufficient guard to protect the
> contents. When they were called away our own people acted as sentries. About the stockade,
> after the vaults had cooled, we placed rough board counters and so resumed business.

Temporary quarters were established in Oakland, and no doubt consideration was given
to the company's long-range future. Both the retail location and factory were rebuilt,

and the workshops were back in production by May, 1908, for they had just then completed a large silver service for the cruiser *California*. The grand reopening in 1909 of the retail store took place almost three years to the day after the first opening on Post Street. Taking advantage of the insurance settlement and the need to retool, Shreve also began development of their first line of flatware. Designed by Joseph E. Birmingham, a salesman with Shreve since 1886, the patterns began to be patented in early 1909. "Louis XVI," "Norman," and "Napoleonic" were the first basic designs, soon followed by "Dolores" and "Marie-Louise" in July. Their most distinctive pattern, "Fourteenth Century," with twelve styles of handles per dozen, was not patented, and within a few years was simplified to one handle design, which in turn seems to have been the inspiration for the "Carmel" pattern by Wallace introduced a few years later.[30]

Equally significant for its recovery and growth, Shreve published the first of its post-Fire illustrated catalogs for the 1909–1910 season. The 1911–1912 catalog, with 224 heavily-illustrated pages, offers an endless variety of jewelry and novelties, as well as their own silverware. Practically everything that could be thought of was offered in silver of their own manufacture, including five flatware patterns and matching holloware in each, as well as an additional eighty pages of imported and luxury goods. Far more eloquently than any cold statistics, these catalogs illustrate Shreve's vast financial and productive resources in the second decade of the twentieth century. Most of Shreve's stylistic innovations had been completed by the time the 1912–13 catalog was published, as was a brief attempt to open a retail outlet in New York. Throughout the 1930s and later, the same designs reappear in countless variations of a few basic patterns and shapes. They continued to do special work and produced items designed by local artists and sculptors, one of which was a silver sculpture presented to Kaiser Wilhelm of Germany, just one year before the outbreak of World War I.[31]

The Shreve factory was quickly converted to war production, and just as quickly converted back to making luxury goods when the war ended. In 1921 their factory occupied 100,000 square feet on six floors, had $150,000 in tools and machinery, and employed more than 600 workers. There seems to have been little design experimentation, however, after their experiences in the Arts and Crafts period. Their "Iris" pattern in holloware is splendid, but makes few departures from what had already been done by Gorham and Whiting. Tastes changed as yet another colonial revival style became popular. For the design of at least two new flatware patterns, Shreve turned to Porter Blanchard, a champion of the colonial style. "Flemish" (1925) and "Old English" (1929) are both based on antique types and have hammered surfaces, produced in the dies.[32]

George R. Shreve, the son of George C., sold his stock in the company in 1912 to the J. E. Hickingbotham family, who in turn sold it in 1967 to Dayton Hudson of Minneapolis. The factory was closed at that time, and Shreve & Company have become again what they were when George C. and Samuel arrived in San Francisco, prestigious retail jewelers who buy their silverware from large Eastern companies.[33]

Shreve's silver production was so large and diverse that any generalizations are subject to many exceptions. Unfortunately, their archives seem to have disappeared, which seriously hampers detailed research. But in the simplest terms, their work can be divided into a few basic categories. First, their special commissions; second, their innovative

Shreve, *"Medieval" Trefoil Dish, c. 1900*

Shreve, *Arts and Crafts-style Bowl, c. 1910*   Shreve, *"Fourteenth Century" Pattern from 1911 Catalog*

flat- and holloware in the medieval revival, Arts and Crafts and art nouveau styles; third, their "colonial revival" flat- and hollowares; fourth, their silver-mounted glass and ceramic wares; and fifth, their rather imitative mass-produced silver of all kinds.

As we have seen, Shreve began their silversmithing with special commissions. This was a reasonable extension of their work as retail jewelers. Their first major works, the Knights Templar trophies of 1883, are remarkable creations (PLATE 8). Large scale, yet meticulously detailed, they make use of a combination of silver, gold-quartz, bronze, and marble to execute complex symbolic designs. Less elaborate, but quite impressive in its own right, is the punch bowl awarded to the Captain of an Alaska steamship by his passengers in 1889, done up in the form of a sailor's ditty bag, complete with a rope chain tie and a pendant anchor at the knot.

Shreve's work in what is loosely called the Arts and Crafts style is a major break from their nineteenth-century traditions. An early and impressive example is the small triangular dish illustrated. The obviously and beautifully hammered surface is pierced with trefoils in a style that sets the stage for many of Shreve's later innovations, including the Medieval style of the massive punch bowl with its complex strapwork, a common feature of both hollow- and flatware of the period (PLATE 12). Most flatware patterns could be ordered either plain or hammered, but a majority of surviving examples are hammered, most likely in the die and not on the piece itself. Another aspect of the Arts and Crafts style can be seen in the "Chicago" style bowl, very much like work from the Kalo Shop. Shreve's art nouveau style is less distinctive and seems to derive from Eastern designs.

The colonial revival wares require little comment, since very similar designs were being produced all over the country, and Shreve's designs do not stand out particularly, except perhaps for their longevity in production on the West Coast.

Among the most interesting of the Shreve productions are their mounted wares. Almost any sort of body, from plain to art glass, pottery and porcelain of English, American, and oriental origin, could be beautifully mounted in silver. Perhaps the most impressive are the rare examples of Moorcroft pottery, sometimes in the form of complete tea sets, set with delicate silver tendrils. The example illustrated is an early piece with a pottery body by the English firm of Taylor, Tunnicliffe & Co., from the turn of the century. Finally, of course, mention must be made of the truly vast quantities of rather ordinary silver which is still more widely owned in San Francisco today than the work of any other manufacturer.

Shreve was also a training ground for silversmiths who later established their own shops and their own reputations. The most important of those who remained in San Francisco were John O. Bellis and Arthur Thumler. Bellis began to work as a jeweler in 1894, and shortly thereafter was employed by Shreve. After 1906 he established his own factory. He was able to attract important commissions from early in his independent career. In 1908 he completed a cup presented by the Chief Petty Officers of the Atlantic Fleet to the "Ladies of the California Club" (PLATE 13). He moved his shop several times, and died in 1943. Bellis's work is characterized by "hammered" surfaces. The basic forms were spun, then meticulously hand-chased. It is not easy to characterize such work as either man- or machine-made. He "hammered" and put his mark on a good deal of flatware and holloware that had been mass-produced plain by various Eastern makers.

W. K. Vanderslice (*left*) *and* John O. Bellis (*right*), *Ewers*

Shreve, *Silver-mounted English Pottery, c. 1890s*

Shreve, *Typical Mass-production Silverware of the 1920s and 1930s*

But he was also capable of making large pieces in traditional designs. One of the illustrated pair of pitchers was made by Vanderslice sometime around 1870, and the other was beautifully made to match by Bellis.[34]

Arthur Thumler, a native San Franciscan, was born in 1886. In 1903 he got a job running errands for Shreve through his older brother Charles, who was employed there as a die-cutter. Arthur soon became an apprentice in the holloware department, where he worked until 1906, when he went East to work for Towle in Newburyport. He returned to San Francisco and to Shreve in 1911. But he also set up a shop in his basement where he executed private commissions for his family and friends. After service in World War I, he did not return to Shreve but decided to set up on his own as a silversmith. By 1920, he had persuaded his brother Charles and younger brother Henry, an engraver, to join his new business. From their shop in the Jewelers' Building on Post Street, Thumler Brothers continued in operation until 1938, when they relocated to 928 Geary in the front part of the space occupied by the Superior Plating Works. Arthur did mainly silver restorations and repair, but he also accepted commissions and made pieces for his family and friends. Arthur retired in 1968 and died the next year.[35]

Arthur Thumler's work is scarce, but is of outstanding quality. Completely competent as a craftsman, he was equally sure of himself as a designer. His designs are sophisticated contemporary interpretations of colonial and early American forms. His fine water-pitcher (PLATE 15) is based on the early-nineteenth-century Liverpool jug form that also inspired Paul Revere, but it is entirely contemporary in spirit. Unlike many of his contemporaries, who preferred to use the hammered surface instead of meticulous craftsmanship, Thumler sought and achieved mirror-smooth surfaces. A family tradition is that Thumler would spend his free time studying traditional forms at San Francisco museums, and he owned a scholarly library of some size on antique silver. He also made coffee pots, candelabra and chalices, as well as small objects such as bracelets, napkin rings, and shakers. His work, which rarely comes on the market, is usually marked "STERLING," "Thumler," and bears a conjoined "AT" touchmark.

Arthur Thumler's death in 1969 symbolically ended the era of Silver in the Golden State. The last "graduate" of Shreve's factory, Thumler had also initiated the study of the history of San Francisco silversmiths. His work, too, fittingly represents the skill and beauty of the San Francisco tradition in silversmithing. California began as a colony of the Spanish; long after statehood, it remained an economic colony of the East coast; but certainly by 1940 California's "colonial" status had ended. When Shreve closed its factory in 1968 it ended the career of California's oldest silversmith and the last major regional silversmith in the United States. From the late-nineteenth-century, production of silverware had increasingly become concentrated on the East Coast. But as the other regional firms closed, Shreve expanded. The reasons for their success are complex. From early in the firm's history it had excellent connections with Eastern suppliers, the management was always socially well-connected, the Company either had or could get ample financing, and by selling only through their own store, they reduced the cost of distribution and kept in close touch with the tastes and desires of their customers. Equally important, they had excellent designers and silversmiths who were given considerable creative freedom, at least in the early years. Their success over more than eighty-eight years gives ample testimony to their skill as merchants and silversmiths. It is encouraging to note that the last few years have seen a small-scale revival of silver-smithing in the San Francisco Bay Area, with a number of well-trained and enthusiastic craftspeople ready to take up once again the long tradition of silversmithing in the Golden State.

(Editor's note: There are, of course, notable California Silversmiths of the post-1940 era, whose work and careers are beyond the scope of this essay.)

Arthur Thumler, *Candelabrum (one of a pair), c. 1950s*

Plates

COLOR PHOTOGRAPHS BY STEVEN RAHN*

*With the exception of PLATES 2 and 9.

PLATE I

(Unknown Maker): *Writing Set* (Mexico City, c. 1800). The tradition of silver as a
sign of rank and wealth was brought to California by the Spanish. This set originally
belonged to Don José Antonio de le Guerra y Noriega (1779–1858), fourth comman-
dante of the Presidio at Santa Barbara.

PLATE 2

HENRI PANELON: *Don Vicente Lugo* (oil on canvas, c. 1835–45). In an exuberant display of wealth and status, Don Vicente's jacket, trousers, and harness are adorned with silver.

PLATE 3

FREDERICK REICHEL: *Pitcher* (c. 1857). The design is renaissance revival, popular in this country in Reichel's day. Although the casting of the handle is reminiscent of Gorham, the shape of the body is Reichel's own design.

PLATE 4

GORHAM CO.: *The Baker Silver* (1860). Presentation pieces commissioned by John W. Tucker on behalf of the merchants of San Francisco, both this urn (and a companion pitcher) exploit railroad motifs in an unambiguous message to Senator E. D. Baker urging his support for the construction of a transcontinental railroad.

PLATE 5

These spikes were driven into a tie of California laurelwood at the "joining of the rails" of the Transcontinental Railroad at Promontory, Utah, on May 10, 1869. The maker and engraver of the silver spike are unknown; the gold spike was made at the William T. Garrett Foundry, and engraved by Schulz & Fischer.

PLATE 6

VANDERSLICE & CO.: *Centerpiece* (with gold wash in bowls, c. 1860–70). This piece shows both the mid-eighteenth century English penchant for convertibility (the bowl-supports are also candleholders) and the mid-nineteenth century American partiality for the machine (notice the gear forms at the piece's base).

PLATE 7

SCHULZ & FISCHER: *Holy Water Bucket* (1880). Ecclesiastical silver, like church ritual, retained traditional forms; the late-Baroque style of this piece reaffirms those traditions rather than setting any new fashions.

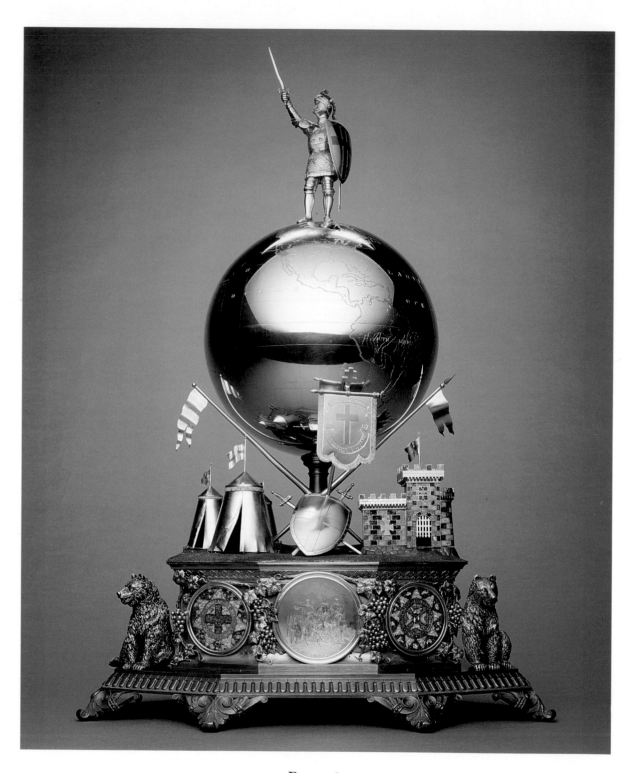

PLATE 8

GEORGE C. SHREVE & CO.: *Trophy* (silver, gold, gold quartz ore, other ore specimens, and bronze, 1883). Commissioned by the Triennial Conclave of the Knights Templar (held in San Francisco in 1883), this trophy was awarded for close-order marching to an Illinois branch of the Knights Templar.

JOSEPH HARRINGTON: *Discovery of the Comstock* (oil on canvas, c. 1875). By 1875, only six years after it was discovered, the Comstock and its discoverers were already being mythologized.

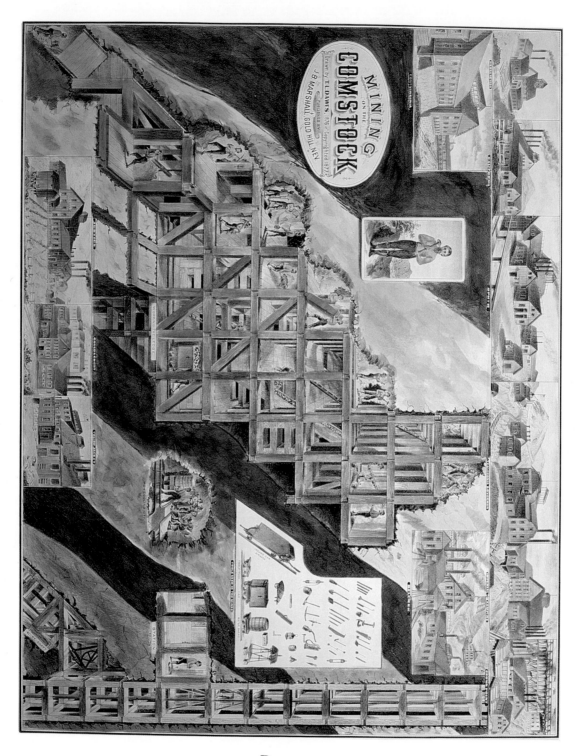

PLATE 10

T. L. DAWES: *Mining on the Comstock* (hand-colored lithograph, 1876). Phillip Deidesheimer's innovative square-set timbering of Comstock ore bodies is illustrated in this lithograph.

PLATE II

SHREVE & CO.: *Plaque* (silver and redwood, 1907). The "drive-through" redwood tree in the Mariposa Grove of Yosemite National Park was a natural subject for a forest products award at a California industrial exposition.

PLATE 12

SHREVE & CO.: *Punch Bowl* (c. 1910–20). A response to the Arts and Crafts esthetic, the hammered surface of Shreve's "XIV Century" pattern suggested handwork, and its applied strap-work had roots in medieval design.

PLATE 13

JOHN O. BELLIS: *Urn* (1908). This presentation piece was given by the Chief Petty Officers of the visiting Atlantic Fleet to the "Ladies of the California Club" in gratitude for a soirée.

PLATE 14

CLEMENS FRIEDELL: *Tea Set* (c. 1920–30). The repoussé-work lavished on these pieces
is typical of Friedell's exceptional talent in that discipline.

PLATE 15

ARTHUR THUMLER: *Pitcher* (c. 1950). Mirror-smooth finishes distinguish Thumler's work, as do his re-interpretations of early American forms (in this case, a Liverpool jug).

PLATE 16

PORTER BLANCHARD: *Tea Set* (c. 1930–35). The clean lines of Blanchard's "Common-wealth" pattern illustrate his predilection for form over ornament.

Arts and Crafts Silversmiths:
Friedell and Blanchard in Southern California

LESLIE GREENE BOWMAN

SILVER PRODUCED in Southern California from 1900 to 1940 is predominantly the work of a small number of silversmiths working primarily by hand. In an age when the overwhelming majority of American silver was mass-produced in the East, the popularity of hand-wrought silver was directly attributable to the Arts and Crafts movement.

The movement was based on the ideas of two Englishmen, philosopher John Ruskin and designer William Morris. These two men believed that the preservation and development of hand crafts provided a socially therapeutic alternative to the monotony of machine-made goods, and indeed even to the erosion of individuality they feared from capitalism itself. Arts and Crafts enjoyed a vogue in the United States during the early years of this century. Although they largely rejected its political aspects, in potteries, print shops, furniture manufactories, architects' offices, craftsmen's associations—and on silversmiths' workbenchs—American artisans adopted the Arts and Crafts tenets of quality materials, sound craftsmanship, and honest, unpretentious design.

California, then and now a seedbed of experimentation, proved fertile ground for the movement. Southern California provided a congenial climate to our subjects—the silversmiths Clemens Friedell and Porter Blanchard—for two reasons: first, it has always been astoundingly tolerant to new ideas, and second, there was sufficient home-grown and transplanted wealth in the area to underwrite the pricey exercises of a silversmith.[1]

It is genuinely ironic that the same movement that informed the creative spirit of both men—Arts and Crafts—and the same social climate and natural landscape that nurtured them both—California—could support such disparate careers.

Clemens Friedell[2] was born in 1872 in Gretna, Louisiana, not far from New Orleans. In 1875 his parents returned the family to their native Vienna, where Clemens was educated. Besides apprenticing with a silversmith for at least seven years, he also studied music. In 1892 the family was back in the United States, and purchased a small farm near San Antonio, Texas. Unable at first to find work as a silversmith, Clemens taught music until 1901. That year, he got a job with the Gorham Company, and with his new bride

Clemens Friedell

Porter Blanchard

Jeanette (née Marcee) he moved to Providence, Rhode Island, the site of Gorham's plant.

In response to the Arts and Crafts movement, Gorham had recently launched a line of hand-wrought silverware, Martelé (French for "hammered"). An expensive line intended for the carriage trade, the Arts and Crafts movement influenced the construction of Martelé but not its art nouveau design. The organic, meandering motifs of this style required skilled repoussé and chasing. (Repoussé is raised decoration, chasing embossed.) Friedell was most likely hired as a Martelé chaser, because a contemporary photograph shows him seated among twenty-five chasers working at the Gorham factory benches.[3]

Friedell left Gorham in 1908; on the back of the photograph of the Gorham chasers he wrote: "started with the Gorham Mf Co. in 1901—left in 1908—opened a shop for myself, operated it for six months—Panic (1907) no work—left for Texas, took up farming—Went to San Antonio after that to California went into Business and still am—this is 1938—Clemens Friedell."[4]

Friedell arrived at Los Angeles in 1910. He lived at first from consignment sales of ashtrays and trinkets at the Broadway, a local department store. Learning of the wealthy residents and visitors of the resort town of Pasadena twelve miles to the north, he moved there the next year.

Pasadena in 1911 was a posh wintering spot for wealthy Easterners. Many built mansions along Orange Grove Avenue for their stay during the "season," while others rented bungalows and suites at the four grand resort hotels or one of the many smaller ones. The Pasadena social season began just after Christmas with the Tournament of Roses, and the next three months were taken up with soirées, balls, polo matches, and excursions on the Pacific Electric railroad to poppy fields, orange groves, missions, Santa Monica, and Santa Catalina Island (by ferry connection).

Friedell rented a cottage, and worked at home. His big break seems to have come as the result of a commission from an important social figure, possibly Phoebe Hearst; a 1912 newspaper article noted that, "[w]hen he came west one of his first commissions was the artistic loving cup, several feet high, for Miss Phoebe Hearst."[5] Whoever the customer was, Friedell's reputation was established, and the carriage trade followed.

In 1912, he received the biggest commission of his career. Los Angeles brewing millionaire E. R. Maier "had heard of the remarkable repoussé work done by the Pasadenan," and ordered an entire dinner service in silver as a birthday gift for his mother. Friedell responded with his masterwork.

He designed the entire service of 107 pieces, which included eighteen each of dinner, soup, salad, and bread and buter plates, along with assorted serving bowls, salts and peppers, tureens, trays, and a twenty-eight-inch-tall centerpiece. Friedell chased the service with more than 10,000 orange blossoms, which was therefore christened "the orange blossom set." He completed the project in just over a year, and for 7,500 hours of work and 2,200 ounces of silver Maier paid him $15,000.

The service is Martelé in style. Arts and Crafts influences are found in the entirely handcrafted character of the set, from the hammered finish to the thin strap handles; "[o]ne hundred and seven pieces of plate were hand hammered . . . , even the handles of the soup tureens and other articles being entirely hand wrought."[6] Friedell's particu-

Equine portrait plaque

Part of the Maier service

Friedell's workshop, 1912

larity for hand work was not uncommon among Arts and Crafts silversmiths, some of whom defined "hand made" as excluding even ancient casting methods long employed in pre-industrial silver. Friedell designed all his own work, generally using heavy sheet stock of seventeen- or eighteen-gauge. He had an assistant, Alf Ellsworth, to raise and prepare work for chasing, although Friedell himself would raise forms when necessary. Spun forms were usually subcontracted, with Friedell completing the decoration; cast and engraved decorations were also subcontracted, although both are rare in his work. Above all, Friedell was a skilled and nearly obsessive chaser (PLATE 14), for whom fourteen-hour days were common. The mark he adopted in 1911, "STERLING FRIEDELL PASADENA" remained unchanged throughout his career.

Friedell prospered. A 1914 advertisement in the Tournament of Roses program shows him holding the Maier centerpiece, with a preliminary drawing of the piece on the wall behind; the caption reads "Clemens Friedell Silver Art Studio—Repoussé & Hand Wrought Silver—Artistic, Original, Individual Designs."[7] A 1918 ad shows an unusual departure from his predilection for the art nouveau—a goblet, pitcher, and tray in colonial revival style, and devoid of any decoration at all. Friedell may either have been appealing to a middle class clientele unable to afford extensively decorated holloware, or it may have been an excursion into a purer style of Arts and Crafts design than what he normally practiced; strict adherants to the movement's ideals spurned excessive ornamentation in favor of simple, well-made forms.

An even purer Arts and Crafts vocabulary is found in Friedell's equine portrait plaques. Designed for Pasadena's well-heeled equestrian aristocracy, these hammered likenesses were fastened to oaken shields with faceted silver nails similar to hardware on Craftsman furniture. The hammered finish was altered to accentuate and complement the various subjects, and many of the plaques hung from heavy, hand-wrought silver chains.

Friedell won a gold medal at the Panama-California Exhibition at San Diego in 1915; unfortunately, the exposition's programs do not show the items he entered, including a huge punch bowl, a coffee set, and several plaques.[8] His entry to the Panama-Pacific Exposition in San Francisco the same year seems to have been limited to a plaque and trophy for the Shooting Association; an avid rifleman, Friedell may have donated the pieces. Trophies in general seem to have been a prominent component of his business. At the 1914 Tournament of Roses, he donated the cup awarded for "best decorated four-passenger automobile."[9] He was also commissioned to make prizes by other donors, and by the Tournament itself. Only a single trophy remains at Tournament headquarters, but a 1914 photograph of his workshop shows numerous trophies, both vases and loving cups, scattered about.

In 1916 Friedell rented retail space in the swank new "pergola" shops at the Hotel Maryland, one of Pasadena's four grand resorts. The shops were built into the existing pergola of flowers and vines adjoining the hotel. A 1917 Pasadena journalist called them "the poetry of shopkeeping," in which "are treasures rich and varied, precious gems, books in deluxe editions, sumptuous furs, [and] treasures of the Orient. Here are works of the silversmith, fine laces, rare china, a florist's shop, a cigar shop, and a miniature stock exchange."[10]

Because he was so successful at the Maryland, in 1921 Friedell was able to retire to Texas and buy a ranch. When his marriage ended in divorce in 1927, Friedell returned to Pasadena and set up shop in the corner of an Oriental imports shop on Colorado Boulevard, Pasadena's main street. He married one of the import shop saleswomen, Eva Woodard, in 1928, and in 1929 moved to a shop of his own at 626 East Colorado.

Friedell weathered the '29 crash and Great Depression without severe difficulty. Enough of Pasadena's wealthy remained to maintain a demand for his work. During this period he completed what he considered one of his best pieces, an ornate cocktail shaker. Friedell's son recalled later that the shaker was a commission from a wealthy man who wanted something risqué. The customer got risqué. The octagonal pitcher has reliefs of nude women on its sides, a nude woman as a handle, cloven feet, a leering Mephistos for a spout, and the lid is four seated female nudes, arms about each other, their curling tresses intertwined, and their posteriors hanging over the lid's edge—from hence the piece's title, "Bottoms Up!" The original commission netted Friedell $5,000, and must have been executed around 1920. Directly after its completion, he started one for himself into which he scratched the date "1920–1936."

Its ill-proportions and crass exuberance aside, the piece does have great technical merit. It is a total departure from the Martelé style. The decoration is dominated by elaborate castings, unusual in Friedell's work. The basic form is Oriental, the chased female nudes are Beaux Arts, and the handle of an arching nude with flowing hair and backstretched arms is indebted to the work of René Lalique. Interpreted not artistically but socially, the piece epitomizes the luxury-loving resort society of 1920's and 1930's Southern California.

The shaker also is illustrative of Friedell's adaptation to changing public tastes. A photograph of his shop from 1935–40 shows not only his typical Martelé work, but also neoclassical pitchers, large chargers and platters with plain or simple gadrooned borders, and lobed-style tea sets and bowls. Friedell changed not only his designs, but the type of items he produced. Prior to his retirement to Texas, special commissions formed the bulk of his work; photographs of his shop show only a few ready-made goods for sale. In contrast, his Colorado Boulevard shop displayed a broad selection of candlesticks, candelabra, platters, bowls, serving pieces, tea sets, goblets, vases, boxes, picture frames, jewelry, and dresser sets. Pasadena's heyday as a winter playground for millionaires was waning, and Friedell could no longer rely entirely on commissions to support himself.

Friedell's business remained lucrative throughout his career, but in a 1960 interview he regretted that the market no longer sustained "great silver artists. . . . Nowadays they tell you to get it out as fast and as cheaply as possible. No artist can work that way."[11] When he died in 1963, no successor carried on after him; he had trained none.

The story of Friedell's colleague, Porter Blanchard (1886–1973), differs from Friedell's in nearly every aspect. Blanchard was American-trained, and his style was simple, clean, and unadorned. While Friedell operated a one-man shop and relied on retail business from a local clientele, Blanchard had as many as twenty-five employees, and conducted his wholesale business on a national scale.

Porter George Blanchard was born in Gardiner, Massachusetts, in 1886, and learned silversmithing from his father, George Porter Blanchard. In 1909 the elder Blanchard

Friedell (in hat) *with "Bottoms Up!"*

Friedell's retail shop

47

opened a shop specializing in flatware, in which he was joined by both his sons, Porter and Richard. All three were recorded as members of the Society of Arts and Crafts of Boston in 1914 (Porter was elected a Medalist, the Society's highest rank, in 1944), and as early as 1912 Porter had exhibited at the Detroit Society of Arts and Crafts, which he joined in 1916.

Around 1914, Porter took over as proprietor of the Blanchard shop, the same year he also married Elizabeth Flood. In 1923, he left the shop and drove across the country to California. Shortly after his arrival, he published his impressions of California in an article, "An Easterner Comes West," in which—in somewhat idiosyncratic syntax—he described himself as a "rank easterner" and:

> . . . one of the first million in Southern California. . . . We're here sure enough, and "no foolin' " in the much advertised and criticized (advertised East—criticized West) land of sunshine and sand storms (last not advertised). That's why we're here (and the other 999,999 also I believe) to see if what you talked about, your wonderful California, was in fact or fiction.

Of his occupation he said, "I am a 'spoon maker,' a silversmith by trade, born and brought up in the service of silversmithing, an arts craftsman from Boston, Mass. . . ."[12]

Like Friedell and thousands of others, Blanchard's arrival was at least partially attributable to the Land-of-Oz image of California propagandized throughout the country by travel agents and real estate developers. At the close of his article Blanchard enthused that "[w]hen California smiles, all is forgiven." In his advertising brochure from 1925, Blanchard did a little polishing of the California image himself: "Sunshine and flowers surround [our] shop; scented breezes, tempered by the Pacific, blow through it. It has expanded wonderfully under this influence; its products have won fame again in the 'West'."[13] (Blanchard was probably more sensitive to sunshine and tempered, scented breezes than most people; he was a practicing nudist, who often did his silversmithing clad only in a work apron. In light of this, how much California's climate may have influenced his relocation to the Golden State presents an interesting question, which, alas, is beyond the scope of this essay.)

Blanchard settled in Burbank, a small town in the San Fernando Valley north of Los Angeles, because "it is cheaper than quarters in larger cities." By 1925 his father and brother had joined him, and a newspaper article reported eight men at the shop "engaged in making a high grade of handmade silverware."[14] Although the shop appeared as "Blanchard Brothers" in a 1926 city directory, Porter was clearly the proprietor, and the shop's mark was his. (His earliest mark on silver, "PB" in an oval, was replaced no later than 1933 by the silhouette mark which he used for the balance of his career; the mark is still used by the firm.) Occasionally some of the shop's pieces would bear an extra "R" for his brother Richard, or the initials of some other workman.

Porter Blanchard was a great hammersmith who enjoyed the raising process; even spouts and handles were usually wrought. He spurned excessive decoration; chasing is minimal on his work, and castings and engraving rare (PLATE 16). Although he preferred holloware to flatware, he produced both in large quantities. Like Friedell, he designed and built his pieces from heavy sheet stock, typically seventeen to twenty gauge. Unless a client specified otherwise, Blanchard polished his pieces to a matte or satin finish; the high sheen seen on most of them today has resulted from years of subsequent polishings.

Blanchard's style was colonial revival. His early work was derived from eighteenth- and early nineteenth-century Anglo-American pieces, and his patterns had names like "John Alden," "Paul Revere," "Colonial Antique," "Mayflower," "Fiddle," "Old English," and "Georgian." A 1925 advertisement listed "Old Silver Duplicated" as among Blanchard's skills. A tea and coffee service purchased from the Detroit Society of Arts and Crafts in 1923 is among the earliest datable Blanchard pieces. The service shows that Blanchard's style was derivative, not imitative; he used an eighteenth-century design vocabulary of baluster shapes, gently domed lids, button finials, molded reel feet, and C-scroll handles, but in a strictly twentieth-century manner. The set demonstrates one of his basic design strategies: in his pursuit of pure, unadorned form, he frequently reinterpreted historical pieces without their chased or engraved decoration.

What decoration did occur in Blanchard's work was usually integral to the form, as in the lobed, organic shapes of the Melon set dating from the 1930's, or the vertical ribbing in a 1925 coffee service in the Old English pattern. In the latter, the design is neoclassical from the eighteenth century, but during that period the vertical ribbing only occurred on drum-shaped tea caddies; the repetition of the design on the tray resulted in a new interpretive form. Similarly, his Lotus pattern combined the convex paneling of the Melon pattern with the neoclassical vase forms of his Old English.

One of Blanchard's design mainstays was Colonial Antique, which he used for tea and coffee services. Patterned closely on historic antecedents, Blanchard nonetheless entirely redesigned the coffeepot; he dispensed with the period "lighthouse" or baluster shapes, and instead vertically expanded the standard apple-shaped teapot.

By 1930, Blanchard had augmented his colonial revival repertoire with more contemporary designs. He modernized traditional forms, and added new Bauhaus-inspired patterns. A large coffee urn from this period is a streamlined version of a traditional form. Blanchard's "Turkish" coffee service is equally simplified, and provides a striking contrast to Friedell's highly decorated 1914 version. The Commonwealth coffee service from 1930 has clean, cylindrical shapes associated with the Bauhaus. (Ironically, the design of Blanchard's Commonwealth flatware pattern is often erroneously attributed to Kem Weber because of its bold, uninterrupted lines.) The transition to twentieth-century design was effortless for Blanchard because he had always emphasized line and form over decoration.

Blanchard's style was permeated by Arts and Crafts ideas. He explained his feelings on the subject in a newspaper article dating from approximately 1930:

> In the days of Cellini . . . design was largely a matter of decoration, and inferior craftsmanship of later days came to conceal careless workmanship by ornamentation. Finally in the days of Ruskin and his followers there was a rebellion among distinguished artisans against this form of rococo ornamentation and out of this came a growing appreciation for simpler, finer lines and plainer surfaces. It is to this school of design that I belong.[15]

In addition to establishing his studio, immediately after his arrival in California Blanchard set about organizing the Arts and Crafts movement in the West. By May of

"Melon" coffee pot

The "Colonial Antique" pattern

The "Lotus" pattern

The "Old English" pattern

1924, he had assisted in—and probably masterminded—the organization of the Arts and Crafts Society of Southern California. Founded to "stimulate the love for the Crafts, to increase the number of Craft workers and to provide a place in which to exhibit and sell craftwork,"[16] the Society numbered more than a hundred members at its first meeting. They made "their first stunt and bow to the public . . . at the Hollywood Fiesta on June 28. In one booth Porter Blanchard, late of Boston Arts and Crafts Society, showed the process of making handmade silver."[17]

Its first president, Blanchard explained the Society's ideology:

> One can only hope for the highest excellence in the smaller things, and happiness. . . . It's just possible it's to be found in the things with which we surround ourselves, the work we know best, the duties that lie nearest regardless of how hunble they may be. The Society of Arts and Crafts of Southern California is behind this thought. It believes in the infinite value of its handicrafts as true examples of careful, loving design and workmanship, well done and well worthwhile for their own value, and the influence for the better they undoubtedly have on other industry.

Blanchard's vision was suitably Ruskinian. He defined craft as "that thing in industry which is good and beautiful and worth doing for its own sake as well as the money made from it," and the craftsman as "a man who has been trained in the practice of some craft with a 'home influence' of love, honor, and respect. . . ." He noted especially that "Arts and Crafts is a constructive educational force."[18] In his view of the machine, however, Blanchard parted company with the English progenitors of the movement. Blanchard and his colleagues felt that the use of machinery was often necessary, and was justified when guided by a hand with "imagination, intuition, integrity, and idealism."[19] He routinely spun holloware on chucks of his own design, reserving handwork for chasing or a hammered finish. Non-circular holloware was hand-raised, and all flatwear was wrought, but polishing was done on a machine wheel.

The Society did not survive past 1930. Fortunately, Blanchard had not relied on it exclusively for his commercial success. He retailed through several shops and studios in Los Angeles, and through Arts and Crafts galleries in other cities. Most importantly, he marketed through department stores, which reached a clientele which might never patronize an arts-and-crafts-oriented shop. Only a year after his arrival in California, Blanchard struck a particularly lucrative deal with the S. & G. Gump Company, a distinguished art and antique firm with stores in San Francisco and Honolulu. With the exception of Los Angeles, where he supplied numerous retailers, Blanchard seems to have limited himself to one outlet in other cities. A 1933 newspaper article remarked on his exports to "almost every large city in America. Only one shop, and that usually the most exclusive in the particular city, is permitted to display his wares."[20]

Predictably, Blanchard's heyday pre-dated the Depression, when he enjoyed the combined marketing potential of various arts and crafts societies as well as large retailers. During the 1920s, he employed as many as twenty-five men in the Burbank studio. Although his wholesale business was predominant, Blanchard maintained a retail shop at the studio. Among his retail customers, he focused on screen stars, undoubtedly for the potential publicity they might bring him.

The coffee pot is a Blanchard piece

Bauhaus-inspired coffee urn

Flatware in the "Commonwealth" pattern

Flatware in the "Georgian" pattern designed for Joan Crawford

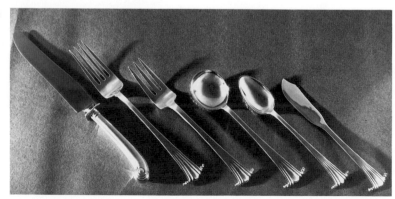

He seems to have done well among movie people. A news clipping reported that "[i]n the homes of those of the film colony who are distinguished by their refinement and love of things beautiful will be found many Porter Blanchard creations. . . ."[21] In 1933, Blanchard opened a retail shop at 6605 Sunset Boulevard, not far from the major studios. An article some years later said that he wanted "to be near folks who had enough money to indulge in such a luxury as high class silverware, especially during hard times."[22] They were hard times indeed: at the time he opened the Hollywood shop, he had already had to lay off employees, rent out his house, and move his family into the Burbank studio. But matters improved with time, and in 1936 he moved the shop west on Sunset Boulevard to a more prestigious location among the stars' agents; he seems to have become the silversmith to the stars. His famous customers included Anne Harding (for whom he designed the Commonwealth flatware pattern), George Brent, Cary Grant, Richard Dix, Joan Bennett, and Joan Crawford, who ordered flatware in the Georgian pattern. He made a great deal of silver for Crawford, and later wrote that "Joan Crawford was probably my best customer and friend although I never saw her face to face ..."; he communicated with the secluded star entirely by telephone.[23]

Porter George Blanchard was an active and practicing silversmith until he died in 1973. His firm still exists, which speaks to his skill as a businessman, to his ability to adapt to changing times and styles. The fact that the firm still produces patterns he created a half century ago argues for their status as twentieth-century design classics.

Both Friedell and Blanchard were drawn to California by the opportunities the state offered to them, whether artistic or financial. But there the similarity between them ends. Friedell was a traditional craftsman trained in the old world, whose style was largely dictated by his penchant for ornate chasing. Blanchard was trained in the United States, and favored simple, unadorned design, which facilitated his adaptation to changing styles in the twentieth century. Friedell catered to the carriage trade, Blanchard to the general retail public. When Friedell died, he left no successor; three years before his death he had acknowledged that the art of chasing was gone except as a hobby.[24] Blanchard left a firm still manufacturing silverwares he originated.

Despite these profound contrasts between them, despite their vastly different styles, Friedell and Blanchard shared a belief in the movement that affected the standards of their craftsmanship. For all the influence Arts and Crafts has had on style, it was not a style at all, but a philosophy that quality and integrity should pervade both the workmanship and design of an artisan's creations. Because Clemens Friedell and Porter Blanchard never lost sight of that, California, in Blanchard's words, smiled on them both.

C. C. Kuchel, *Virginia City 1861*

Bonanza!

The Comstock Lode and California

DONALD L. HARDESTY

THE COMSTOCK LODE of western Nevada was the most famous and important source of California silver.[1] Between its discovery in 1859, and its decline by the early 1880s, over $330 million was extracted from the Comstock, more than half in silver; expressed at today's prices, that figure translates into many billion of dollars, a sum that had a profound effect on California and the nation.

Nationally, Comstock bullion was a critical source of financing for the Union during the Civil War. In California, the Comstock catapulted San Francisco from a port town of middling status to the most important city on the west coast of North America. Fabulous personal fortunes were made on the Lode; those of Flood, Fair, Mackay, O'Brien, Sutro, and Hearst are only the most famous. Besides Nob Hill mansions for the "Bonanza Kings," Comstock money in San Francisco built the Palace Hotel, the California Theater, the Bank of California building, the Union Block, the Nevada Block, the Flood Building, and the Sutro Baths. The Lode also proved rich in the lore of the American West; Mark Twain, by way of example, cut his literary teeth on the pages of Virginia City's *Territorial Enterprise*, and his time on the Comstock is a cornerstone of *Roughing It*.

For all the fame that attaches to the Comstock, it is still little known that the book on hardrock mining was rewritten there. The geological, technological, and legal problems posed by Comstock mining spawned innovative solutions; "square-set" timbering, flat braided hoisting cable, the safety cage, large-scale and efficient exploitation of steam power, huge ventilation and drainage schemes, and the Washoe milling process first made their appearance on the Lode. National mining law was created and refined there; one critical development allowed an ore body to be followed underground wherever it went, irrespective of the above-ground borders of the claim. Miners' unions played a key role in the development of industrialized mining. And the Comstock had a rich social and ethnic diversity. These are among the lesser-known facets of Comstock Lode history to be explored below.

The Lode was discovered in 1859 on the slopes of what is now Mount Davidson in the Virginia Range of the Sierra Nevada mountains. It was created by the injection of hot mineralized fluids from the earth's interior into two major and several minor faults during the late Miocene epoch, about 12–13 million years ago.[2] Large amounts of quartz containing silver and gold were deposited, much of the silver occurring in the form of silver sulfide, especially the mineral Argentite.

The main body of the lode is oriented north–south and dips forty-five degrees toward the east, down the slopes of Mount Davidson. In places it is more than sixty-five feet wide. The first Comstock mines were located near the upper end of the lode, high on the side of the mountain. These "first line" mines struggled with the practical and legal problems of whether the Comstock Lode was made up of one or several ledges, and whether it dipped toward the east or the west. That the Lode was a single ledge and dipped downhill toward the east was clear by 1864. The following fifteen years saw two more "lines" of mines developed farther down the hillside, both lines attempting to locate new and deeper ore bodies. Mines as deep as 3,300 feet had been dug by the 1880s.

The fifty-six major Comstock mines struck several bonanzas or major pockets of silver ore, including six by 1864. Of these, the Ophir Mine at one of the original discovery sites was the most famous. By 1865, however, these mines began to play out, causing a general economic depression on the Lode. It was not until the Crown Point–Belcher bonanza in 1870 that the Comstock enjoyed renewed prosperity. In 1873 the last major bonanza was discovered at the Consolidated Virginia Mine in Virginia City. The "Con Virginia" produced over $64 million of precious metals by 1880,[3] making it the wealthiest of the Comstock mines. No new bonanzas followed, and the Comstock's "silver age" came to an end in the early 1880s. The Comstock mines, however, continued to produce silver throughout the 1880s and 1890s.

Working the great depths of the Comstock lode was made possible by the development of new technology that went on to revolutionize the mining industry. Several key innovations, such as "square set" timbering and flat woven cables transformed Comstock mines from modest human-scale operations into giant industrial enterprises.

The early mines had been worked using cheap hand tools or simple machines generally powered by human or animal labor. By the mid-1860s, however, the Comstock was industrialized, with widespread use of powerful and expensive steam-powered machines. The reason for the transformation was the mining itself; Comstock mines were the deepest in the world, and the mines were dug into an ore body that constantly threatened to collapse. Many of the mines encountered heat and water that created immense engineering problems of drainage and working conditions. A new hardrock mining technology emerged from the solutions to these problems.

Development of the Comstock mines was hindered by the instability of the crushed, clay-filled, and water-soaked character of the ore bodies. None of the timber-support methods then known seemed able to keep the mines open; cave-ins plagued them. In late 1860, however, a timbering system of "square sets" was devised by Philipp Deidesheimer at the Ophir Mine.[4] A German immigrant with recent experience developing gravel mines in the vicinity of Georgetown, California, he used cube-shaped modules of heavy timbers that could be locked together in all directions to completely brace the

Consolidated Virginia Mine 1877

T. L. Dawes, *The Belcher Mine*

mined-out space of the ore body. In this way, large ore bodies could be rapidly and economically mined. Even square sets were inadequate to prevent cave-ins for extended periods; Dan De Quille visited the Consolidated Virginia Mine and vividly described what happened in some places no longer being worked:

> The large, square timbers are crushed down to half their original height and are splintered and twisted; chambers originally square are squeezed into a dimaond shape, and their roofs almost touch the ground in the center. . . .[5]

To slow down the process, the mined-out sets were often filled with waste rock or low grade ore to give additional support.

The tools first used on the Lode were simple: picks, shovels, hammers, rock drills, and black powder. Rock drills were driven into the exposed face of the ore body with hammers, either by single miners ("single jacking") or in teams of two—one holding the drill and one driving ("double jacking"). The resulting hole was filled with black powder and detonated. In 1867 dynamite was invented and soon thereafter used in the mines.[6] "Giant powder," as it was first called, was safer and easier to control than black powder. Another key invention was the Burleigh mechanical drill, invented in 1869 and first used on the Comstock at the Yellow Jacket Mine in 1872.[7] By 1874, it was being used in the Consolidated Virginia and the Gould & Curry mines, and was in general use shortly thereafter.[8] Run by compressed air, the drilling machine eliminated the slow and laborious task of driving rock drills by hand; when combined with yet another invention, the diamond rock drill bit, the Burleigh and its descendants opened a new era in industrial mining.

During the first few years of work on the Comstock, the hoisting of ore, waste rock, supplies—and the miners themselves—into and out of the mines was done in a variety of ways, all relatively simple. The windlass, a hand-lifting device with a reel, a cable, and a large iron bucket, was in general early use. By 1864, they were replaced at many mines by horse-powered whims. At the Mexican Mine rawhide sacks of ore and waste rock were hand-carried with the help of a tump-line. And sometimes a small steam engine was used to pull an ore car up an incline.[9]

New methods were adopted as the mines grew deeper. Large steam engines replaced earlier hoisting machines. In his classic *Comstock Mining and Miners*, Eliot Lord observed that by 1866 the forty-six mining companies then working on the Comstock were using forty-four engines for hoisting and draining the mines.[10] Within fifteen years the size of engines used had increased many times; the Yellow Jacket alone had installed steam engines totaling 2,941 horsepower, including two 1,000 horsepower hoisting engines, nearly double the horsepower of *all* engines used on the Comstock in 1866.[11] The foundry industry in San Francisco and Nevada boomed in response to the massive demands for metal and metal-working created by the machines.

The energy demands of the machinery were enormous; the Comstock burned fuel in prodigious quantities. For example, the Yellow Jacket alone consumed thirty cords of yellow pine every day and 10,950 cords during the year.[12] Several large lumber companies grew up in the nearby Sierra Nevada forests to satisfy the Comstock's appetite for fuel, and entire forests were cut clean within only a few years.

Underground at the Utah Mine

Hoisting Works of the
Con Virginia and California Mines

Carson & Tahoe Lumber and Flume Co.s'
Facility at Spooner's Summit 1876

By the middle 1860s, simple iron frame compartments had replaced the earlier hoisting buckets in most mines. These open "cages" were capable of running up and down a shaft at speeds in excess of 700 feet per minute. Dan De Quille made such a trip in the Consolidated Virginia and compares the experience to riding in a fast train at night:

> Our speed is such that we see but little. We get a glimpse of what appears to be a room of considerable size, see a few men standing about with candles or lanterns in their hands, hear voices, and probably the clank of machinery. An instant after, all is again smooth sailing and we see only the upward-fleeing sides of the shaft. Then there is another flash of many lights, a glimpse of half-naked men, a murmur of voices, and a clash of machinery, and we have passed another station.[13]

The cables hoisting the cages sometimes broke, sending the manned vehicles plummeting to the bottom of the shaft. Two key inventions greatly decreased the danger. The first was the flat woven wire cable which was made originally for the Sierra Nevada Mine by A. S. Hallidie and Company of San Francisco.[14] Flat cable was much stronger and lighter than hemp rope and did not slip as much on the hoisting reel, allowing more control in operation. The other invention was the safety cage. First installed at the Savage Mine, this cage had a safety device made up of heavy springs that would propel a toothed wheel into the wooden rails at the side of the shaft if the cable suddenly broke.[15]

Despite these improvements, however, riding the cages was still dangerous. The cages were not at first enclosed, and many miners fell off to be crushed between the cage and the shaft. The precise, demanding work of the hoist operator sometimes took its toll; after long hours of work, the operators made mistakes such as allowing the cage to be drawn into the hoisting reel, throwing the miners out into the shaft or the machinery. Public outcry over such accidents was largely responsible for reducing the work shift of the operators to eight hours a day.

One of the fundamental problems in mining is maintaining adequate ventilation of the underground workings. Both poor air and heat created problems in the Comstock mines. The air was so bad at times that candles carried by the miners burned at only one-fifth their intensity in normal air. The heat was so intense in some places that miners had to strip to breechcloths, shoes, and hats, and consumed enormous amounts of water and ice to keep cool enough to work; Eliot Lord observed that in the hottest parts of the Consolidated Virginia Mine during the summer of 1878, each miner consumed an average of three gallons of water and used ninety-five pounds of ice during an eight-hour shift![16]

The problem of ventilation was approached in two ways, either by forcing air into the mine, or by creating an air "draft." In the early 1860s outside air was forced into Comstock mines by several simple methods. Dan De Quille mentions the use of wind-sails (large cloth bags that filled with air when turned to the wind and forced air through a tube into the shaft), blacksmiths' bellows, and water blasts (forcing air through a tube by pouring water into it).[17] By 1865, forced-air ventilation could be greatly improved by the installation of large rotary blowers or fans, such as the Root blower.[18] The compressed air systems used to run the Burleigh mechanical drill often were used as a power source for such blowers.

"Draft" ventilation works with air as a siphon does with water. A shaft was sunk from the surface to intercept the underground workings at some point distant from where they met the main hoisting shaft. The warm, lighter air from the mine interior would rise out of one shaft, sucking cooler, heavier fresh air from the surface down the other shaft, and through the tunnels and passageways of the mine below. Of the ventilation systems used on the Comstock, the draft method was by far the most efficient. In the Bullion Mine in 1877 it was estimated that about 300,000 cubic feet of air flowed out of the underground workings every minute, of which 10,000 came from air compressors and 30,000 came from blowers; the rest was draft ventilation.[19] The system was slow in coming to the Comstock mines, however; early mines were developed with little advance planning, resulting in isolated workings that were difficult to interconnect. But by the 1870s and early 1880s draft ventilation was fairly widespread.

In the vast deserts of the American West little is more important than water. It is not surprising, then, to find that the Comstock was preoccupied with water—ironically, both too much and too little. Lord observed that flooding historically has been the bane of the miner's existence.[20] The Comstock was no exception. In the beginning the mines were usually drained by digging horizontal tunnels ("adits") into the mountainside to intercept flooded shafts. The 1,100-foot-long "Union Tunnel" was dug in 1860 to drain the Mexican, California, Central, and Ophir Mines to a depth of 200 feet,[21] and Lord mentions ten other adits that had been dug by 1863.[22] Most drainage problems in the early years were effectively solved this way, with the addition of small pumps powered by steam engines and by bailing the shafts with the iron buckets used for hoisting. But as the depth of the Comstock mines increased, the drainage problems became acute. A solution proposed by Adolph Sutro was to dig another adit, but this time on a grand scale; the tunnel Sutro proposed was over three miles long, and would intercept the lode at a depth of 1,663 feet. The Sutro Tunnel project was approved by the state of Nevada in 1865 and begun in 1869, but was not completed until 1878 when the Comstock was well into decline.

The most practical solution to the drainage problems at most mines was the use of ever larger and more powerful pumps. The first steam-powered pump on the Comstock, of a modest fifteen horsepower, was installed at the Ophir Mine in late 1859 or early 1860.[23] In 1861 then Superintendent Philipp Deidesheimer recommended that a larger "Cornish" pump be installed. Producing forty-five horsepower, the pump nevertheless was thought to be "fearfully extravagant" by the Ophir's directors;[24] nonetheless, within a few years all of the mines were using massive and costly Cornish force-pumps. By 1879, for example, the last such pump to be installed (at the Union shaft) generated 540 horsepower and was capable of pumping 2 million gallons of water a day.[25] The Cornish pumps were supplemented by hydraulic pumps in the early 1880s; working together at the Combination shaft they pumped over five million gallons a day.[26]

Drainage was only one of the water problems on the Comstock; the other was the lack of adequate surface water to run the wet Washoe-milling process, and for human consumption. The mills met the problem in two ways: either by building on the nearby Carson River (the Excelsior Mill) or in Washoe Valley (the Ophir Mill); or by building water collection and storage systems on the Comstock itself (the Gould & Curry Mill

Pump at the Union Shaft

Arrastras at Virginia City 1864

in Six Mile Canyon). Perhaps the most elaborate water engineering effort undertaken was to supply the towns and mills of Virginia City and Gold Hill. In 1873 the Virginia City and Gold Hill Water Company hired Hermann Schussler, chief engineer for the Spring Valley Water Works of San Francisco, to develop the water resources of the Carson Range for its use.[27] Schussler engineered a remarkable system of collection reservoirs, wooden flumes, and pipelines. The key component was an inverted siphon through which water dropped from the Carson Range 1,997 feet to the floor of the Washoe Valley below. At this point, under a tremendous pressure of over 800 pounds per square inch, the water was forced by gravity 1,526 feet upward into a reservoir that fed the Comstock communities.[28]

The Comstock was significant not only for its deep mining technology and water engineering but also for the development of new and efficient ways of recovering silver from its ore. Milling silver ores was a novel proposition for the Comstock miners. Most had come from the California gold fields, where milling was simply crushing ore and collecting the free gold on sluices or broad riffled plates. Only some of the Comstock ores could be processed this way, mainly the earliest to be taken from the Gold Hill mines. Most of the ores were silver compounds, not metallic silver, and much more difficult to mill. At first, Comstock ore was sacked and shipped to England for milling. Eventually, "pan amalgamation" was adapted from milling technology developed in the silver mines of Mexico and Peru. Here, the ore is finely crushed and mixed with mercury, salt, and water, and then heated for several hours in a large pan or kettle. The process causes a chemical reaction that forms an "amalgam" of silver and mercury. Heating the amalgam vaporizes the mercury (which is condensed and recovered), leaving the metallic silver behind.

On the Comstock the "patio" process of pan amalgamation was used first; the patio combined the Mexican "arrastra" grinding technique with solar-heated amalgamation. Arrastras were flat, circular stone floors upon which the ore was piled and then pulverized by placing large flat rocks on it and pulling them around a central pivot with animals or water-power. Mercury, salt, water, and copper sulfate were added to the finely ground ore and blended, often by men or animals walking through the mixture (or "pulp"). The pulp was then spread on a rock "patio"; exposure to the sun's heat created the amalgam. In Mexico, the patio method was a widely used, cheap (albeit labor-intensive), way of milling silver ores on a small scale. On the Comstock it turned out to be inefficient, in large part because sunlight was not adequate to heat the pulp properly for much of the year.

After a two-year period of experimentation, the early Comstock mills developed what became known as the "Washoe process" of pan amalgamation. The Washoe process combined the arrastra and the patio in a single large iron pan. Silver ores from the mines were first crushed in large stamp mills, something like mechanical mortars and pestles. The standard "California" stamp mills commonly used steam engines to turn a cam shaft upon which were mounted several heavy iron posts or "stamps," which pounded ore placed in a mortar box below; large mills might have as many as eighty stamps. After crushing, the ores were shoveled into large iron pans about four feet in diameter, holding

*Batteries at the Mill of
the Gould & Curry Mine*

California Pan Mill 1882

California Pan Mill - 1882.

a "charge" of about 300 pounds of ore, salt, mercury, water, and copper sulphate. The ore was then ground under a heavy rotating iron plate or "muller." The pans were equipped to heat the pulp, either through the direct injection of steam into the pulp or a steam-heated chamber beneath the pan. With this artificial heat, amalgamation could be completed in several hours rather than several days. The amalgam was recovered by passing the pulp into a series of "separators" and "agitation" tanks, where the heavier silver-mercury compound settled out and was removed. In the final operation, the amalgam was heated in retort furnaces; the mercury, which evaporates at a lower temperature than silver, was driven off, leaving silver bullion.

The first successful mill operator on the Comstock was Almarin B. Paul, who set up a 24-stamp mill and used an early version of pan amalgamation in the fall of 1860.[29] By the end of 1861 the *San Francisco Evening Bulletin* reported seventy-six mills on the Comstock.[30] Many of the early mills were elaborate showpieces, perhaps intended to impress potential investors as much as anything else. The Gould & Curry mill, for example, was a large "hacienda," with an elegant mansion-like building, elaborate landscaping, water reservoirs, outbuildings, domestic animals, and workers' cottages.[31] Later mills, however, were much more austere and efficient, reflecting the economic realities of Comstock mining.

Prior to the Comstock's discovery in 1859, the first to explore the Comstock were prospectors.[33] Mostly adult males, they usually worked alone or with a partner. They were of diverse origins, with substantial numbers of Mexican and Chinese miners among a variety of Anglo-American and European nationalities. The prospectors lived in isolated cabins or huts rather than in camps; Johntown, a small camp in Gold Canyon settled in 1855, was an exception.

By and large, the prospectors were an egalitarian lot, with few significant differences in wealth or property. The reason is that Comstock prospectors of the 1850s mined placer gravels. Placer mining promotes social "leveling" in a number of ways. First, the free gold in placer gravels was usually evenly distributed in small amounts, making it unlikely that one miner would accumulate much more than another. Secondly, the shallow placers could be worked with cheap hand tools (such as long toms and rockers) available to everyone; there was no possibility for social differences to emerge because of one or a few miners gaining control of the "means of production."

The development of the Comstock mines became more and more expensive over time; raising necessary capital was always a pressing problem. At first, mines were owned by individuals or small partnerships capable of mustering only relatively small amounts of money. Mining corporations were soon organized to increase available capital. The Ophir Mine, for example, which was worked by an association of owners in 1859, was incorporated as the Ophir Mining Company on April 28, 1860.[32] Shares were offered to the public. The Comstock mining companies raised working capital through assessments levied upon shareholders, who lost their shares if they were unable or unwilling to pay. Otherwise, the shares had little relationship to the assets, reserves, and overall fortunes of the mines. The shareholders, for example, had no say in how the mine was operated. And the shares, although at first being something like deeds to a given number of feet of a mining claim, were soon changed into stock certificates bearing no connection to the traditional "claim feet." The vast amounts of capital raised by mine stocks mostly went into the pockets of speculators, rather than into the shafts of Comstock mines. Trading in Comstock mining stocks was a mania in the United States and Europe for nearly 20 years. The San Francisco Stock and Exchange Board was organized on September 1, 1862, specifically for the purpose of trading shares of Comstock mines, and stock exchanges in Virginia City and Gold Hill soon followed.

In 1864, a depression resulting from the playing-out of the first bonanzas radically changed how the mines were financed and controlled. The Bank of California, newly organized in San Francisco by William Ralston and D. O. Mills, sent William Sharon to the Comstock in 1864 to act as its agent. Sharon lent money at a low rate of interest but quickly foreclosed when, because of the depression, his clients defaulted. In 1867 the Union Mill and Mining Company was incorporated by Sharon and the owners of the Bank of California to buy up the foreclosures; the "Bank Crowd," as this group was known, controlled all of the major Comstock mines and mills by 1869.

The monopolization of the Comstock by the Bank Crowd came to an end in the early 1870s. New personalities and corporations emerged, the most important of which were the "Bonanza Firm" and its four "Bonanza Kings." Perhaps the best known of them was John Mackay, an Irish immigrant who came to the Lode from the California gold fields in 1859. Working first as a miner, he became superintendent of several mining corporations, none of which struck bonanzas. In 1865, however, Mackay and a partner invested heavily in the Kentuck Mine in Gold Hill. The Kentuck struck a large ore body in 1866, making Mackay quite well off; but Mackay's 1869 involvement with the Bonanza Firm at the Hale and Norcross Mine made him rich. The group included Mackay, James Fair, James Flood, and William O'Brien. Fair was a mine superintendent who came to the Comstock in 1865; Flood and O'Brien were San Francisco saloon-owners-turned-stock-brokers who never actually lived on the Comstock. The new group acquired other mills and mines. Besides the Hale and Norcross, in 1871 the Firm gained control of the Consolidated Virginia Mine, which struck the Comstock's largest bonanza in 1873. The era of the Bonanza Kings ended about 1880, when the "Con Virginia" paid its last dividend.

The Comstock strike was followed by a population explosion that turned Gold Canyon and Virginia City into an urban strip. Virginia City was the largest of the towns, with a size of over 20,000 in the 1870s. All together, perhaps 30,000 people lived on the Comstock during its heyday.

Residents of the Comstock lived in an exciting world. Dan De Quille said this about Virginia City:

> During summer, men who have for sale all manners of quack nostrums, men with all kinds of notions for sale, street-shows, beggars, singers, men with electrical machines, apparatus for testing the strength of the lungs, and a thousand other similar things, flock to Virginia City. Of evenings, when the torches of these parties of peddlers, showmen, and quack doctors are all lighted and all are in full cry, a great fair seems to be under headway in the principal street of the town—there is a perfect Babel of cries and harangues.[34]

They also enjoyed many of the luxuries found in much larger nineteenth-century cities, including opera houses, hotels, restaurants, and a host of goods and services—the Comstock traded in a global market place. Clothing, food, furniture, and other commodities were imported from California, the Eastern United States, Europe, and elsewhere. Restaurant menus of the time reflect the somewhat extravagant urbanity of Comstock life; the 1878 Christmas menu of the International Hotel in Virginia City, for example, lists a variety of French and California wines, oysters on the half-shell, mock turtle soup, salmon, trout, leg of mutton, ribs of beef, Westphalia ham, canvasback duck, quail, venison, crab salad, roast turkey, capon, stewed terrapin, a large number of vegetables, and an enormous number of pastries and desserts. The Comstock also drew heavily upon local resources to supply its needs; in the late 1870s there were eleven dairies in and around Virginia City, cattle were raised in the Truckee Meadows, fruits and vegetables in the Carson Valley, and game taken from the Sierra Nevada foothills.[35]

Residents of the Comstock lived in a highly ordered society. Eliot Lord was one of the first to observe that the social order in Virginia City was visible in where the residents lived on the slopes of Mount Davidson.[36] The wealthy lived on the upper streets; below this were the main business districts and homes of the working class; further down the slope at the bottom of the social ladder could be found the red light district, Virginia City's Chinatown, and the scattered dwellings of Native Americans. The disparity in wealth between "upper street" Comstock society and everyone else was enormous. In addition to the "Bank Crowd" and the "Bonanza Kings," many of the mine super-intendents and leading merchants became millionaires. Some of their mansions still stand on the slopes above C Street, Virginia City's main street, remainders of the upper reaches of the Lode's geographic social hierarchy.

Theoretically, Comstock society was "upwardly mobile," and the industrious or fortunate could become wealthy. In reality, considerable pressures prevented more than a few from ever doing so.

Comstock miners were among the most admired of nineteenth-century workers, in part because of the danger of working in the deep mines. They commanded the highest mining wages in the world, at least double the pay of any other mining district. Most were American, Canadian, English (mainly from Cornwall), or Irish. Significant numbers also came from Germany, Scotland, Wales, Nova Scotia, and Sweden.[37] The federal censuses for Virginia City show that Mexican miners were on the Comstock in reasonably large numbers until 1870, but had disappeared by 1880. Chinese hardrock miners are conspicuously absent, largely because of the exclusionary policies of the miners' unions.

Comstock miners were described as "well fed, well clothed, and well lodged."[38] At the same time, they clearly occupied a second-class position on the Comstock social scale. Many lived in Gold Hill, considered a "working class" town. Those who lived in Virginia City had houses not on the upper slopes with mine and mill owners and managers, but farther down toward the center of town, where they lived in rather drab, unpainted wooden frame cabins or lodging houses. The miners did mix with the Comstock elite in some ways, however. The best and most expensive restaurants in town were visited by miners and owners alike. Miners shared with the Comstock *crème de la crème* a love of playing the stock market. Miner and nabob alike were entertained at the "opera houses";

International Hotel, *Virginia City*

Edgington Residence, *Virginia City*

Virginia City Miners

the spectacle of bulldogs fighting bears, or of Adah Isaacs Menken disrobing, must have had nearly universal appeal.[39] Plays, recitals, lectures, and other more sedate diversions were also democratically patronized.

Any purported glamour of the miner's life pales in the light of his working conditions. Comstock mines had heat and foul air conditions that were never adequately abated and which became worse as the mines were dug deeper. Temperatures of over 100 degrees were common, even after the installation of large blowers in the late 1860s. The miners often had to strip to thin overalls or a breechcloth, shoes, and hat to work, and he consumed massive amounts of water and ice to cool off.[40] Shifts that required working as little as fifteen minutes every hour were used in some of the deepest and hottest shafts, and miners working there were paid a higher wage.[41] Despite precautions, miners died of heat exhaustion and suffocation, although this was not a major cause of mining fatalities. The water that plagued the mines posed another hazard. By 1880 water temperatures as high as 170 degrees were encountered in the deeper workings.[42] A number of miners were scalded to death by falling into pools of hot water. Falls down mine shafts, however, were the most important cause of mining fatalities on the Comstock; a miner plummeted down the shaft, tearing off limbs on protrusions from the walls on the way down—between that and impact with the bottom, the remains later recovered were little more than an unrecognizable mass. Other accidents were just as grisly. Miners were sometimes crushed between shafts and the rapidly moving hoisting cages, until enclosed cages were installed. Some were killed by underground fires, such as at the Yellow Jacket and Crown Point mines in 1869, and in cave-ins. The frequent observation that the Comstock miners were "fatalistic" is therefore scarcely surprising.

With few exceptions, the miners were unable to do much about either the working conditions in the mines, or gaining a more equitable share of the Comstock's wealth from the few who controlled it. Their primary accomplishment was to fight for a minimum wage and an eight-hour shift by organizing the first miners' unions west of the Mississippi. In 1864, with the Comstock in economic depression, a "Miners' League of Storey County" was formed to assure that no miner would be paid less than $4 a day.[43] Miners paraded through the streets and to mines paying lower wages. Local newspapers, such as the *Territorial Enterprise* and the *Gold Hill News*, took up their cause. Another campaign started in 1867 with the formation of the Virginia City Miners' Union. The articles of organization stated that miners should be paid no less than $4 for an eight-hour shift. In 1869 the miners convened the "Workingmen's Convention" at Virginia City to insist that no Chinese laborers be hired by the mining companies. This demand sprang partly from fear that Chinese miners would work for wages less than $4 per shift, and partly from simple racism. By 1872, the miners had succeeded in unionizing Comstock mines.

Bonanza Comstock society was cosmopolitan. In 1875, for example, the social calendar of Virginia City includes festivals and rituals of Germans, Italians, French, Chinese, Swiss, Cornish and English, Irish, Mexicans, Scots, Jews, Chileans, and Canadians.[44] All of these and more gave a color to the Comstock communities usually found only in much larger cities. But the very visible physical and cultural differences of several minorities

Chinese Quarter, Virginia City 1870's

Indian "Campoodies" below Virginia City

Virginia City High School Class of 1883

forced them into ethnic enclaves and restricted the jobs and lifestyles available to them. In Virginia City's geographically arranged social order, they were the ones living at the bottom of Mount Davidson; of these, the Chinese immigrants and Native Americans were the worst off.

Several hundred to as many as 2,000 people lived in the Virginia City Chinatown. Except from a few contemporary photographs, not much is known about the community. Newspaper accounts, diaries, and other primary source material are biased, giving only a warped view of them.[45] Nevertheless, the Comstock Chinatowns appear to have been typical of the Chinese "sojourner" communities found elsewhere in the West.[46] For the most part coming to America with the intention of later returning to their homeland— rather than becoming permanent residents—they attempted to retain traditional ways of life; they lived together, spoke their own languages, dressed in traditional ways, and ate traditional foods when possible. The immigrant Chinese were mostly adult males who had left their families behind in China. They often lived in what Westerners perceived as crowded quarters. On the Comstock, the sojourners worked as day laborers, ran laundries and restaurants, cooked, and hawked firewood. They were willing to work long hours for wages much less than most other ethnic groups, and the mining unions would not allow them to work in the mines for this reason. They did work placer gravels, however.[47]

Native Americans whose traditional homeland included the Virginia Range continued to live on the Comstock after its invasion by the miners. Some Washoe or Paiute settlements, for example, were found at the very bottom of Mount Davidson, next to the mine dumps and mill tailings below Virginia City. The traditional lifestyle of these people was greatly altered. Foraging for plants and game in a seasonal cycle was largely replaced by wage labor, especially on ranches. Other jobs available to them included laundress, maid, kitchen helper, water carrier, construction laborer, guide, and ore assorter.[48] Some foraging was continued; the traditional fall forays to the piñon pine forests to gather pine nuts, for example, were often mentioned in newspapers. Foraging for clothes, shelter materials, and food took place in Virginia City itself, a new twist on an old practice.[49] The Comstock Native Americans adopted western clothing and used western materials to build their traditional domed "wickiup" or "campoodie" shelters, which now became pipe frames covered with canvas.

Other ethnic groups were present, but little is known of them with any certainty. Virginia City had a community of about ninety blacks, most of whom lived downtown on "C" (the main) Street.[50] What is known of them is limited to information in federal census records and a few photographs. From these, we know that they owned small businesses, including barber shops and boot-blacking shops, and worked as teamsters, laborers, and in a variety of other jobs.

Cerro Gordo

Cerro Gordo

DONALD CHAPUT

CERRO GORDO translates from the Spanish as "Fat Mountain," and, mineralogically at least, the Inyo County peak lived up to its name. For a decade it was second only to Virginia City as a silver producer, and second to none in the production of lead. While no California mines could hope to challenge the preeminence of Nevada's Comstock Lode, the mining and smelting of Cerro Gordo silver, its shipment to San Francisco via Los Angeles, and the people responsible for it all, make a story worth telling. A Mexican, a Yankee, and a pair of French Canadians are the principal *dramatis personae*. First onstage is the Mexican.

Although the boundary between fact and legend is blurred, it seems that early in 1865 one Pablo Flores and a pair of his countrymen were the first non-Indians to visit Cerro Gordo. There on a prospecting excursion from Nevada, they were promptly captured by a band of Pah-utes. Having extracted from them their promise to leave and never return (under pain of what punishment, we do not know), the Indians released the three Mexicans. Despite his pledge, that autumn Flores returned with other Mexicans to prospect for the silver veins he had spotted earlier in the year.[1] In the first months of 1866, José Ochoa located a claim he called the San Lucas, and began the first systematic mining of silver on the mountain. By April 5, 1866, Cerro Gordo was a going concern and the Lone Pine Mining District, encompassing the workings, was organized.[2]

Newspapers began to report bits and snatches of intelligence from the District. The *Mining & Scientific Press* of San Francisco described several of the new silver operations, and the Los Angeles *News* reported on crudely processed specimens of Cerro Gordo silver ores appearing in that city.[3] These and other published accounts sparked a small silver rush to Cerro Gordo.

By 1867, the District had matured to a point that correspondents could report that: a town had been laid out, many buildings erected, and the settlement would be "permanent and prosperous";[4] that there was much payable ore; that Ochoa and crew were doing nicely at the San Lucas; that other major workings included the San Pascual,

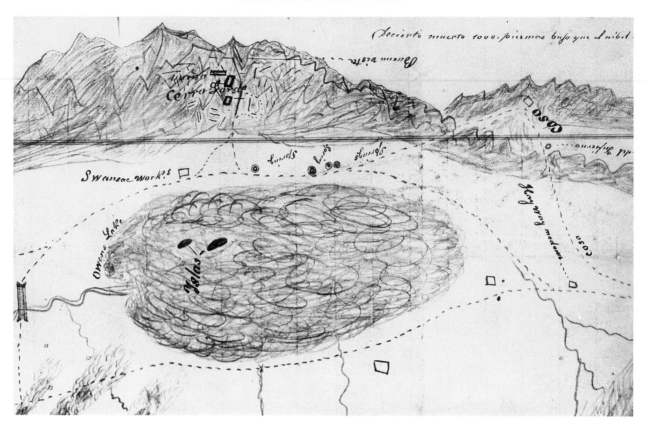

*Manuscript Map of Cerro Gordo
by Manuel Coronel*

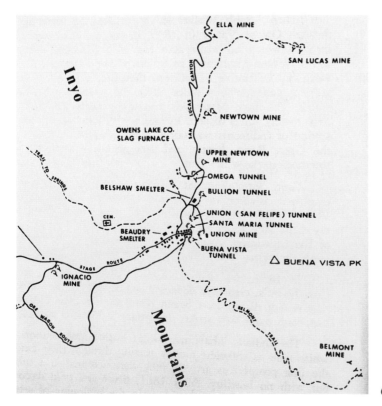

Cerro Gordo, 1869–79

Santa Maria, Buena Vista ("an immense vein"), Union, Belmont, and Bismarck; that, taken all around, Cerro Gordo displayed every sign of a "flourishing mining country."[5] It was also noted that the silver ore was being smelted at numerous *vasos* at the mines, at a *patio* operation at Owens Lake, and at a pan process mill at Kearsarge, some forty miles distant.[6] Bear this latter fact in mind; we will soon see that the decentralized smelting operations and the crudity of the *vaso* process would drastically alter the who's who of Cerro Gordo.

For the present, however, admirable descriptions of the *patio* and pan may be found in Dr. Hardesty's article elsewhere in this volume. The *vasos* evolved at silver mines in Mexico more than a century and a half before. Stone and adobe ovens with low ceilings to reflect heat downward, when fired to high temperature they simply roasted silver out of its ores. Mexican miners at Cerro Gordo occasionally added a preliminary roasting, a modest refinement of the process. Stacking alternate layers of pine logs and ore, they burned the whole affair for several hours, driving off much of the waste material in the ore as gas. The remaining smoldering mass was shoveled into a *vaso* to be refined "in due Aztec style" to bullion.[7] To refer to the "crudity" of the *vaso* process is to do it no injustice; 1708 accounts of *vasos* in Chihuahua show them to have been virtually identical to those used by Mexican miners at Cerro Gordo in the 1860's.[8]

As news of doings at Cerro Gordo spread in the press and along the miners' grapevine, Americans and Europeans began to appear on the mountainside in significant numbers, and for a time it even looked like Cerro Gordo might become a satellite of the Comstock. A prospector named Charlie Duval and several other Virginia City emigrants filed early claims at Cerro Gordo, and kept Nevada newspapers abreast of developments. Duval wrote the *Virginia City Trespass* in September, 1867, that "the Virginians are delighted with their ledges and prospects" at Cerro Gordo.[9] Other Virginia City papers for the same month mention homes and smelting furnaces erected by Nevadans, and that the prospects for the Lone Pine District were of "a very flattering character."[10]

A particularly sanguine Nevada emigré and Cerro Gordo booster was Dr. James Delavan. A '49er from New York, Delavan was an old and experienced Comstock hand, lately superintendent of both the Mount Davidson Tunnel and the Lady Bryan Mine, and known and respected in Virginia City mining circles. He arrived at Cerro Gordo in November, 1867, to manage the newly formed Cerro Gordo Mining Company. Impressed with what he saw, he reported it to be the best mining prospect on the West Coast, although he also noted the District lacked capital, an assayer, and an adequate transportation system.[11]

He also recognized the one really glaring deficiency—the lack of proper smelting facilities. The *vasos* scattered across the mountainside simply were neither efficient nor economical.[12] To correct the problem, Delavan had his own smelting furnace in operation by early 1868; unfortunately, it was only a slight improvement on the status quo. Hoping to produce a really proper smelter, he continued to puff the District and urge investment by Comstock capitalists, claiming Cerro Gordo would soon be "the most busy mining region in the world."[13] But Virginia City turned a deaf ear to him; with hundreds of shills and con artists promoting marginal mining ventures all over the West, her investors had heard Delavan's brand of hyperbole before. They certainly had enough to preoccupy them at home; this was the era of the first bonanzas on the Comstock, and Cerro Gordo—no matter how "flattering" its prospects—would have looked like so much small change in comparison.

Despite the failure of his furnace Delavan remained optimistic, and for a time continued to mine, smelt, and run assays in Virginia City. As late as May of 1868 he had arranged a shipment of supplies and equipment from Nevada. But for all his best efforts, the end of his particular enterprise was near; Delavan's last reports from Cerro Gordo were under date of that same month. In April he had noted a fact that would ultimately foreclose any substantial investment at Cerro Gordo by Comstock interests: "Parties have already arrived in the camp from San Francisco, have secured mines, and are about erecting furnaces."[14] Delavan, José Ochoa, and Pablo Flores were about to be elbowed aside by a pair of men who, financed by San Francisco capital, would dominate "Fat Mountain" for the next decade.

Although they had little in common, Victor Beaudry and Mortimer Belshaw collaborated in a handsomely profitable ten years on Cerro Gordo. Beaudry, a native of Quebec, came from a prominent merchant family. He and his brother Prudent arrived in California in 1849, opening businesses in San Francisco and Los Angeles. During the Civil War, Victor was a sutler (a private merchant of food and drink) to Union troops. When the War ended, he moved to the new garrison town of Camp Independence in the Owens Valley to open a store. In 1866 he was one of the first merchants to set up at Cerro Gordo.[15]

A recent commentator has noted that Beaudry established himself on the mountain "in accordance with the callous business customs of the day." He extended credit to Mexican miners, and when they couldn't make payments attached their property; he got hold of the Union Mine, a prime operation, from José Ochoa this way.[16] Even though Beaudry was cozening indigent Mexicans in a remote frontier mining hamlet, it would be unfair to him to view him simply as a small-time gouger swindling innocent locals; Beaudry was definitely a big-time gouger—he merely started small. He had money, experience, and connections. During Beaudry's stay at Cerro Gordo, his uncle was the leading merchant of Montreal, where his brother Jean-Louis was mayor. Brother Prudent was at the time mayor of Los Angeles.[17] While sutling during the War, Beaudry's operations had taken him at various times to San Francisco, Los Angeles, St. Louis, and New York, his business a far cry from that of the usual sutler trundling along behind the troops with a cart or two of merchandise.

Although he had moved to the Owens Valley as a merchant, intending to provision the army outpost at Camp Independence, he was not ignorant of mining. Over the years Beaudry had been an officer of or owned considerable shares in gold and silver mines in the San Gabriel Valley.[18] In 1858 he had been treasurer of the Santa Anita Mining Company, a gold mining concern near Los Angeles. In 1861 he was president of the Azuza Mining District and wrote its laws regulating silver mining. He was in a position to appreciate the augeries presented by the first operations at Cerro Gordo, and set to work acting on them. From Ochoa, by fair means or foul, he acquired controlling interests not only in the Union Mine but also the San Lucas; he and a partner built a pair of smelting furnaces, whereupon Beaudry promptly elbowed out the partner.[19] By early

Victor Beaudry

Mortimer Belshaw

Workings of the Union Mine

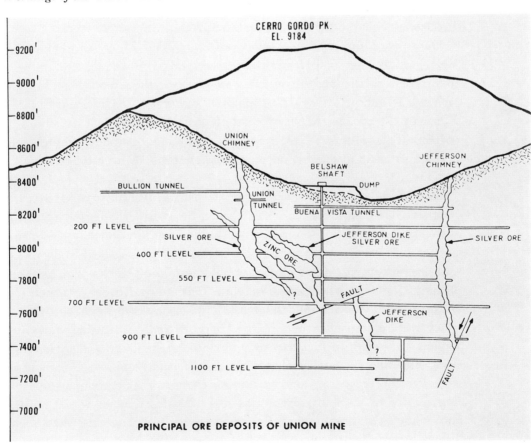

PRINCIPAL ORE DEPOSITS OF UNION MINE

1868, Beaudry had two stores, two furnaces, and control of the two best mines on the mountain.

Mortimer W. Belshaw arrived on the mountain in April, 1868, having read news of the Lone Pine District in San Francisco papers. A native of New York, Belshaw had trained as a watchmaker and jeweler and done a stint as a collector on the Erie Canal before emigrating to California in 1852. He settled in Fiddletown, Amador County, where he was the Wells Fargo agent and had a jewelry store.[20]

Belshaw was reported in Sinaloa, Mexico, from 1862 to 1864, apparently to operate a silver mine in which he had invested. While this fact cannot be established with certainty, it is plausible; there were at the time a number of important silver mines in Sinaloa—the Cinco Señores, the Copala, the Cosala—many of them backed by San Francisco capital. A later source suggested that Belshaw left Mexico "owing to the instability of the government," abandoning his claim before it had been developed to any significant degree.[21]

Settling in San Francisco, Belshaw established the Pacific Refinery, smelting gold and silver ore with only marginal financial success. Perhaps that galled him; this was also the time of the first great Comstock bonanzas. Or perhaps some of the newspaper accounts through which he learned of Cerro Gordo were Dr. Delavan's, bemoaning the lack of capital and expertise necessary to exploit the district's mines.[22] Whatever the exact cause, with an eye to the Comstock's example, his experience in silver mining and smelting, and nothing more than an unprofitable refinery keeping him in San Francisco, Belshaw decided to try his luck on the Fat Mountain.

He was accompanied to Cerro Gordo by Abner Elder, another veteran of Mexican silver mines. They inspected the workings, talked to the locals, and were impressed by one fact in particular; the miners were working the silver veins, but the Union Mine, rich in lead deposits, was attracting little attention. Belshaw and Elder knew from their Mexican experience that lead is an efficient flux for smelting silver. They saw that, instead of just mining silver, they could make a lot of money refining it. Belshaw bought a third interest in the Union, mined a few tons of ore, and rented a nearby *vaso* to process it.[23]

In June, a wagon loaded with ingots, Belshaw, and Elder rolled into Los Angeles. The partners accompanied their cargo aboard the steamer *Orizaba* to San Francisco, where it was refined and the resulting silver bullion used by Belshaw as an enticement to potential investors.[24] He secured the backing of one Egbert Judson, president of the California Paper Company and a dabbler in mining stocks. The two formed the Union Mining Company and, capital in hand, Belshaw and Elder returned to Cerro Gordo.[25]

It must have been about this time that Belshaw and Beaudry struck their own deal. They were already partners in the Union mine. With capital now available to fully exploit Cerro Gordo, Belshaw may have found Beaudry's stranglehold on the District appealing, and Beaudry probably thought Belshaw's prowess at smelting (as well as his access to capital) of no small usefulness. Perhaps they recognized in each other kindred spirits. Beaudry's sharpness in business has already been noted; neither was Belshaw any piker as an opportunist. At one time, he had converted the sole road into Cerro Gordo into a private toll road. When a court order subsequently put the kabosh on the

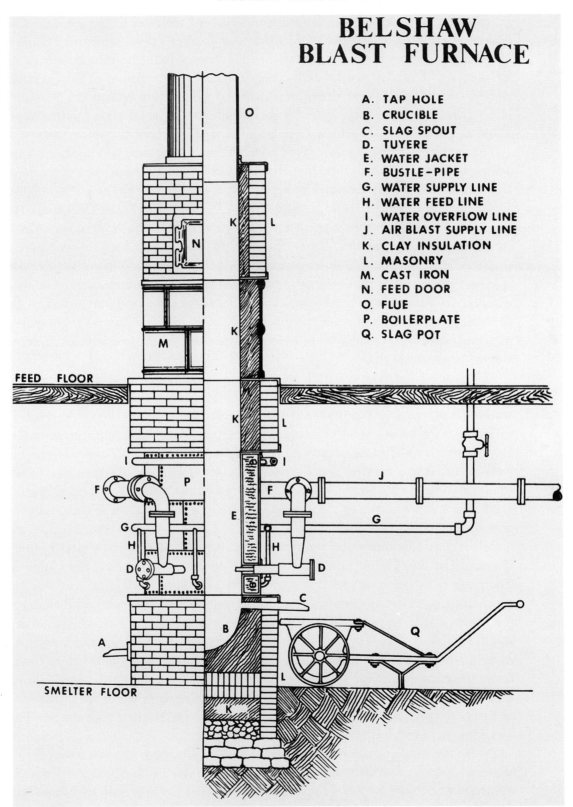

BELSHAW BLAST FURNACE

A. TAP HOLE
B. CRUCIBLE
C. SLAG SPOUT
D. TUYERE
E. WATER JACKET
F. BUSTLE-PIPE
G. WATER SUPPLY LINE
H. WATER FEED LINE
I. WATER OVERFLOW LINE
J. AIR BLAST SUPPLY LINE
K. CLAY INSULATION
L. MASONRY
M. CAST IRON
N. FEED DOOR
O. FLUE
P. BOILERPLATE
Q. SLAG POT

FEED FLOOR

SMELTER FLOOR

Belshaw's Refining Furnace

81

scheme, Belshaw simply walked away and allowed the road to deteriorate. The *Inyo Independent* noted with asperity that Belshaw "makes a poor return to a country that has yielded him hundreds of thousands of dollars; very, indeed." Whatever their reasons, Belshaw and Beaudry shook hands, at least figuratively, and joined forces. In doing so they avoided the expensive and acrimonious competition that often plagued other mining districts, and between them were able to more or less dictate how matters would stand at Cerro Gordo for the next decade.

The logistical problems they faced were ferocious. The first was location. The mines at Cerro Gordo were at an elevation of nearly 9,000 feet above sea level, and a mile above the floor of the Owens Valley. The Valley itself (dominated by Owens Lake, described at the time as having a "fetid, sickening odor"), was arid, remote, and sparsely populated. Wood and water were needed in vast quantities for both human and industrial consumption. The wood was cut in forests on the east scarp of the Sierra Nevada some thirty miles away, floated down streams to the floor of the Owens Valley, and then hauled eight tortuous miles over the steep road from the valley to Cerro Gordo. Water was piped in from several miles away in a system financed by investors from Los Angeles; the inauguration of the system was marred by a failure of the pumps, to the general dismay and disgust of both residents and investors. Once finally in operation, however, the system performed reliably for years.[26]

The town of Cerro Gordo itself was problematic. Like other western mining camps, life there could be stormy. Beside the majority of largely law-abiding families and workingmen, there was the obligatory contingent of drunkards, crooks, and hell-raisers. Contemporary accounts tally the typical inventory of mining camp recreations: drunkenness, brawls, fandangos, prostitution, gunfights, stabbings, robberies. The town was in *bandido* territory, making travel and transport hazardous, dangers exacerbated by Mexican miners who, when they were displaced by the influx of Anglos, joined up with established highwaymen in the area like Tiburcio Vásquez and Cleovaro Chavez (who once robbed Beaudry himself).[27]

Belshaw and Beaudry also ran into that pandemic affliction of western mining, the lawsuit. The San Felipe Mining Company filed an action against the Union Mining Company (i.e., Belshaw and Beaudry) in 1871, claiming the Union was working lead ores belonging to the San Felipe. This was a serious threat; Belshaw and Beaudry needed the Union's lead to preserve their monopoly on smelting at Cerro Gordo—to lose it would ruin them. Years of claims, counter-claims, threats, and acrimonious squabbling culminated in an 1875 California Supreme Court ruling in the San Felipe Company's favor. But the judgment was largely irrelevant. All through the years of litigation, Belshaw and Beaudry had the courts' permission to continue working the disputed vein, and they simply stripped it of ore before being forced to share it with the San Felipe; it was a loss in court, but a victory underground.[28]

By far the biggest problem facing Belshaw and Beaudry was the transport of their ingots to market. The ultimate destination of Cerro Gordo silver was the precious metal refiners and brokers of San Francisco. But between Cerro Gordo and San Francisco, thirty miles across the Owens Valley, stood the Sierra Nevada. And not just the Sierra Nevada, but the *highest* range of the Sierra Nevada; thirty-two miles due west and exactly opposite Cerro Gordo is Mt. Whitney, at 14,495 feet the tallest peak in the contiguous forty-eight states. The nearest northern passes that could accommodate freight wagons were hundreds of miles away and closed by snow in winter. The only practical alternative for Belshaw and Beaudry was to skirt the mountains to the south, take the ingots to Los Angeles (180 miles distant as the crow flies, 275 as the wagon rolls), and from there ship it by sea to San Francisco. The journey was arduous. In winter it involved long hauls through the arid and uneven terrain of the lower Owens Valley, Red Rock Canyon, the Mojave Desert, and Cahuenga Pass; in summer there was the additional problem of withering, prostrating heat.

Belshaw and Beaudry turned to Remi Nadeau, a Los Angeles teamster, for a solution. A French-Canadian like Beaudry, Nadeau set up relay teams of dozens of wagons and hundreds of mules scattered along the route between the mines and the coast. For several years his wagons carted their seven-ton loads of silver-lead ingots into Los Angeles, their "bells jingling, wood and leather creaking, blacksnake (whip) popping, mule-skinner shouting and cursing, mules snorting and coughing."[29]

In 1871, despite his apparently satisfactory services, Belshaw and Beaudry saw fit not to renew Nadeau's contract; they had been seduced away by competing teamsters. For the next two years the transport of Cerro Gordo bullion was a hit or miss proposition. A brisk municipal competition for the Cerro Gordo trade developed, with Ventura, Visalia, Santa Barbara, and Bakersfield clamoring to edge each other—and especially Los Angeles—out of the picture. Several of them were closer to Cerro Gordo than Los Angeles, and Visalia and Bakersfield could additionally tout prospects of future rail connections to San Francisco. The stakes were high. There was considerable commerce involved in transshipping the bullion north and provisioning and maintaining both Cerro Gordo and its wagon system; it ultimately figured large in the development of the city of Los Angeles. In lamenting its ultimate failure to divert the Cerro Gordo trade to itself, Bakersfield's *Californian* confirmed that fact: "Few comprehend the immensity of the demands such transportation by teams makes upon farmers for supplies. It has been the great traffic of Los Angeles, and has perhaps done more to build up that city than any other one enterprise."[30]

Along with the distracting competition among the towns, the several carriers retained by Belshaw and Beaudry at one time or another during these two years proved unreliable. Compounding their own incompetence or inexperience, these contractors were bedevilled by equipment breakdowns, a severe equine epidemic, and the anticipated railroad's failure to materialize. The result was that bullion bars by the thousands backed up at the Cerro Gordo and Owens Lake staging areas; there were so many at one time that miners took to using them for housing, building igloos of ingot walls and canvas roofs.

With their freight situation—and undoubtedly their cash flow—in a critical bind, Belshaw and Beaudry must have grown nostalgic for the good old days when reliable and capable Nadeau handled their shipping chores. They ultimately worked out a new deal with him, the three forming the Cerro Gordo Freighting Company, with Los Angeles once and for all fixed as the terminus of the wagon route.[31] Nadeau then engineered an even more remarkable scheme than his earlier one. Between Cerro Gordo and Los

One of Nadeau's Teams in the
Mojave Desert

Bullion Backed-up at Owens Lake

Angeles, he established a dozen stations a day's haul apart, each with water, feed, and repair facilities. Eighty teams of fourteen mules pulling three wagons were on the road at any one time; averaging eight round trips a year, the teams were a perpetually moving conveyor belt of bullion going southwest to Los Angeles and provisions northeast to Cerro Gordo. In addition to solving Belshaw's and Beaudry's shipping problems, Nadeau's operation made him one of the wealthiest men in California.

Although the mines at Cerro Gordo had been located and worked for silver, commercial amounts of other minerals, including zinc and gold, were produced. Of these ancillary metals, however, lead was by far the most important. Not only was it the flux they used to smelt the silver ores, Belshaw and Beaudry discovered by happy chance that they had a ready market for the lead itself. Thomas Selby, a '49er from New York, established his Selby Lead Factory at the corner of First and Folsom Streets in San Francisco in 1865 for the production of shot for firearms; his 180-foot-high shot tower was a local landmark.[32] (A shot tower produces shot by dribbling molten lead from a platform at the top; as the lead falls, it forms into round droplets, which solidify as they cool on the way down.) Selby flourished, and was soon the sole shot manufacturer on the West Coast. In 1868 he opened a foundry in the North Beach district of San Francisco for the smelting of lead and manufacture of products such as pipes, vat linings, solder, and paint pigments. Note that 1868 was also the year that Belshaw and Beaudry's operations at Cerro Gordo commenced; it was a serendipitous coincidence for all hands. Selby had been acquiring his raw lead from mines scattered over Utah, Nevada, and the lower Colorado River basin.[33] Because they used lead as flux in their silver smelters, Belshaw's and Beaudry's bullion consisted of ingots of silver and lead, several ounces of silver to roughly eighty pounds of lead, plus a pound or two of other minerals; the total weight of each bar was, on average, eighty-three pounds.[34] When the silver from the ingots was removed in San Francisco, naturally a large quantity of lead was left over. It found a ready purchaser in Selby, undoubtedly grateful for a centralized and reliable source of supply, and provided Belshaw and Beaudry with a tidy windfall for what otherwise might have been waste material. Largely due to Cerro Gordo lead, the Selby foundry became one of the largest producers of lead in the country.

Cerro Gordo's salad days ran out in 1877. Several events combined to hasten the end: a nearby gold strike caused a minor rush, depleting the labor supply at the Cerro Gordo mines as miners left for the new gold workings; a fire at Beaudry's smelter caused extensive damage, curtailing production; the richest silver ores had been removed, and the remainder were only marginally profitable. Most critically, the lead deposits of the Union mine, which had made possible the profitable smelting of Cerro Gordo silver, were by now played out. Contemplating these facts (and probably reflecting on their healthy balance sheet), Belshaw and Beaudry called it quits, folded their tents, and decamped. The more or less total collapse of the Lone Pine Mining District and the town of Cerro Gordo attended their departure.[35] Some sporadic mining for silver, gold, lead, and zinc went on over the next decades, with meager results.

The total income produced by Cerro Gordo mines during their heyday is impossible to determine with certainty, but credible estimates place it at between six to ten million dollars (in 1870's dollars; the value today would naturally be far more). It was nothing, of course, on the order of the Comstock, but an appreciable sum nonetheless.[36]

The Hotel Nadeau

Thomas H. Selby

The principal recipients of that sum put their shares to good use. Mortimer Belshaw betook himself to the town of Antioch in Contra Costa County, somewhat north and east of San Francisco. There he established the Empire Coal Mine, the Belshaw Mercantile Company, the Bank of Antioch, a local water company, and invested heavily in the famous (and profitable) Kennedy Gold Mine at Jackson in Amador County. He died in Antioch in 1898, leaving his various enterprises to his son Charles, a state senator from 1900 to 1919.[37]

After leaving Cerro Gordo, Victor Beaudry went first to Los Angeles, where he acquired the local water works and partnered his brother Prudent in both the largest real estate office in the growing city and on the board of directors of the Temple Street Cable Railway. He retired to Montreal where he died in 1888, leaving an estate estimated at $1,500,000.[38]

Remi Nadeau, too, profited handsomely. Among a dozen interests he acquired in Los Angeles were a winery, a large vineyard, a sugar refinery at Santa Monica, and the Nadeau Hotel, for years a Los Angeles landmark, and when he built it the first four-story building in town and the first with an elevator. He also had the foresight to leave a descendant and namesake, another Remi Nadeau, who glowingly chronicled his ancestor's accomplishments in the book *City-Makers*. The elder Nadeau died in Los Angeles in 1887.[39]

Of Cerro Gordo itself little remains. There are several houses, including Belshaw's, that still stand, although they definitely show their age. There are a few ruins scattered about. None of the mines have been worked for years. The present population is precisely two.

Silver bullets (and a horse to match)

Icons of Continuity and Tradition:
Some Thoughts on Silver and the California Experience

GERALD W. R. WARD

GOLD IS THE precious metal that usually comes to mind when one thinks of California. One's mental picture of the Golden State is dominated by images of the Golden Gate and its Golden Gate Bridge. The gold helmets of the San Francisco '49ers remind us weekly during football season of the Gold Rush of 1849. The official state flower is *escholtzia california*, the Golden Poppy; the state fish is *salma aguabonita*, the Golden Trout; the state colors are Gold and blue; the state mineral is, of course, Gold. Where else but California, one wonders, could the citizen arm him- or herself with a pistol embossed in pure gold (the golden bullets are extra) from the Beverly Hills boutique of designer Bijan Pakzad?[1]

The tremendous impact of gold on California can be verified by even as small an object as a fiddleback teaspoon. Inscribed "Made of native gold by Moffat & C° Sanfrancisco, Calª 1849" on the back of its handle, the spoon is the work of John L. Moffat, a New York goldsmith who emigrated to California in 1849 to establish an assaying and smelting business. Other examples of early California gold pieces survive, including an elaborate seven-piece tea set presented to Mayor Cornelius Kingsland Garrison in 1854 by the citizens of San Francisco.[2] These and other objects bear eloquent witness to the significance of gold to California.

Yet Silver, gold's noble cousin, also pervades the Golden State's lore and imagery. California is the land of the Silver screen, on which one can watch films like the classic *The Silver Streak*. California is home to the Silver and black of the Los Angeles *née* Oakland Raiders. (The Raiders and '49ers between them have won five Super Bowls; California can therefore lay a larger claim to the Lombardi Trophy, the Silver football and kicking tee made by Tiffany & Company, than can any other state.) The Silver bullets of the Lone Ranger, a California-made staple of the early years of television (a mini-Silver screen), are the progenitors of Bijan Pakzad's golden ones. And we all know the name of the Ranger's horse. In San Francisco there still survive some of the "Silver Palaces," the homes and hotels built by the "Silver Kings," the capitalists who made their fortune in the silver mines of the Comstock. Thus, metaphorically at least, *Argent* as well as *Or* can stake a claim to California's metallic imagery.

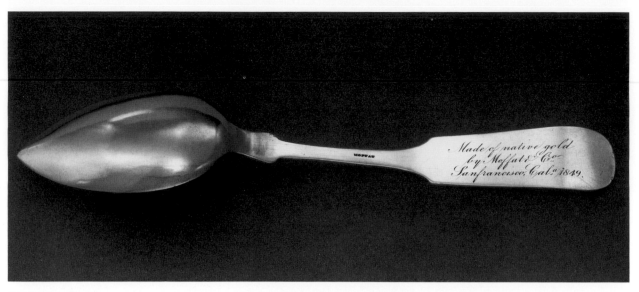

The Moffat gold spoon

Silver Screen, 1949

But silver has played a larger role in the state than merely as a source of figures of speech. Silver in the California experience also involved the metal's tangible and traditional function, in objects made both for everyday use and for special occasions. Although the production and ownership of silver in the state is represented by but a few objects surviving from pre-Gold Rush times, there is a substantial body of artifacts from the second half of the nineteenth century when pioneer firms like Vanderslice & Company and Shreve & Company established the West Coast's version of the silversmithing and plating industry of the East. Our own century is represented by fine objects handcrafted by Allen Adler, Clemens Friedell, Porter Blanchard, and other artisans. Silver in California includes silver owned there but made elsewhere, whether in the United States or abroad, and these imported objects must also be considered in any rounded view of the role silver has played in California life. Even modern collections of antique silver are an important aspect of silver's current presence in the state. Paying particular attention to the symbolic significance silver objects had for their owners, these are the aspects of silver in the California experience which we will explore.[3]

Silver is a magnificently versatile material which can be fashioned into objects which not only reflect the skill of their makers, but which can also tell us much about the attitudes and values of their owners. The physical properties of silver—malleability, ductility, durability, reflectivity—have made it especially suitable for objects given to mark our various rites of passage, literally from the cradle to the grave. Our persisting high regard for the metal has also made it appropriate for use in other rituals, whether religious—such as a holy water bucket made by Schulz & Fischer of San Francisco in the 1880's—or domestic—such as a tea service made of Comstock silver by Gorham in 1874. This tea set did double duty; besides handling the domestic chore of serving tea, it also marked a passage, having been commissioned by the miners of the Belcher and Crown Point Mines as a wedding present for their superintendent. A symbiotic relationship is at work here; the importance of the event (a wedding, a christening, an anniversary) lends stature to the silver object, and in turn the presentation of a silver object confirms the significance of the event. (PLATE 7.)

In modern times, especially since the latter half of the nineteenth century, we find that silver, once a commodity only for the elite, has entered most peoples' lives in ways, both large and small, not always predicated on ownership. Silver has been used to enrich the more prosaic rituals of everyday life, through implements like letter openers, shaving mirrors, and grape shears. The use of silver in coinage, exploiting both its intrinsic value and its durability, is within recent memory, albeit our dimes and quarters are now copper and nickel sandwiches.

For centuries, ownership of silver served as a barometer of social status. This purpose was most effectively served through its use and display in the home; your silver could not impress your friends and guests unless they could see it, and since the seventeenth century silver has traditionally been placed on view when not in use. (Lamentably, this practice has declined in the twentieth century out of fear of theft.) In late seventeenth century New England, for example, silver objects and expensive imported glass and ceramics were displayed atop an oak cupboard in the homes of the well-to-do. Later, high chests of drawers and corner cupboards were used for this purpose, and by Victorian

Child's flatware gift set

The Belcher and Crown Point miners' gift

Trophies—for golf, Peanut League baseball, and yellow-eyed male cats

times the dining room sideboard had become the favored vehicle of display. These side-boards often all but literally groaned with the weight not only of a family's new silver, but with the old silver that towards the end of the last century began to be taken out of storage and valued as antiques.

Such a sideboard, sufficiently laden with bowls, candelabra, trays, and tureens, literally and figuratively reflected a new era in the history of the consumer's relationship with silver. The peak period of settlement in California coincided with a significant shift in the perception of silver objects. Prior to that time, silver could attribute much of its allure to the fact that it was scarce. Although much more abundant than gold, silver was nonetheless rare enough in the seventeenth, eighteenth, and early nineteenth centuries for its possession to be restricted to the economic and social elite. Even huge infusions of the metal into Europe after the Spanish discoveries of silver mines in Mexico and Peru did not widen the circle of ownership of silver objects beyond its traditional patrons in the church and aristocracy. At the time of America's colonization in the seventeenth century, silver still functioned as a signifier of wealth and social status. This role diminished somewhat in the late eighteenth century, as small silver implements and jewelry became more common with the early stages of the Industrial Revolution. By the start of the nineteenth century, however, the ownership of silver was still limited along traditional lines.[4]

Several things happened during the last half of the nineteenth century to funda-mentally change our perception of silver goods. The making of objects, for one, evolved from craft into industry. Labor-saving techniques like drop-stamping, rolling, spinning, and machine engraving supplemented—and largely replaced—traditional methods of handcraftmanship. It became easier to produce larger quantities of objects, and a firm like Vanderslice and Company of San Francisco could employ twenty or thirty men to do so.[5] A second key development was the introduction of electroplating into this country in the 1840's. By depositing only a thin layer of silver onto a base metal, plated objects of almost any size or configuration could be cheaply produced and widely marketed. Although the plating industry was based in New England, there were plating works in California by the 1870's. Haynes and Lawton in San Francisco, for example, were the agents for the Pacific Plate Works, and advertised in 1869 that they would use "a full deposit of silver" on objects of nickel silver or white metal, to create plated ware "equal in Durability & Finish to the imported Goods." The San Francisco Plating Works advertised two years later in a similar vein, announcing that they did replating and carried "a large assortment of new styles of Plated Ware and Cutlery . . . for sale at the lowest rates."[6]

Electroplating had a somewhat pernicious effect on silver's allure. A veneer of silver on a base metal object gave it the appearance but not the substance of sterling. The American fondness for plated goods may have derived not only from their cheapness, but also from an American penchant for what historian Daniel Boorstin has called "the image." A plated teapot is a form of "pseudo-event," not unlike a Hollywood press conference or other contrived occasion. It is not only a form of reality in itself, but is also a simulation of a "real" thing.[7]

The innovations in manufacturing and the introduction of plating were fueled by the discovery in the American West of vast amounts of the raw material needed for these processes—silver itself. The 1859 discovery of the Comstock Lode at Virginia City, Nevada, marked a new era for the metal, as silver came out of the ground in prodigious quantities over the next decades. One writer characterized American silver production in the early nineteenth century as "inconsequential." By 1875 national production exceeded twenty-four million ounces, in 1900 fifty-seven million, and in 1915 was just short of seventy-five million ounces.[8] While most major mines were in Nevada, Colorado, Utah, and Arizona, California mines (such as at Cerro Gordo) contributed their share to the bonanza.[9] In the past, silversmiths largely had to resort to melting down old or broken objects or coins for their raw material; now a vast and available supply was at their disposal.

This combination of circumstances—improved manufacturing technologies, electroplating, and the abundant supply of silver and its concomitant lower price—forever altered the mystique of the metal. Historically, silver objects had been available only to a wealthy few and been made by a small number of craftsmen who enjoyed high standing in the community (witness Paul Revere, for example). Now ownership was far more egalitarian, and the objects themselves made in factories by anonymous workmen and sold through middlemen retailers. One need only contrast the scarcity of silver objects from California's Spanish and Mexican periods with the second half of the nineteenth century to see this point confirmed.

Ironically, the mass production of plated goods stimulated the production of sterling. To genuinely distinguish themselves, the wealthy now needed ever larger and more elaborate services, replete with almost endless arrays of specialized flatware and holloware. An early California example is a service made in the 1850's by Tiffany and Company for James Birch (d.1857), a successful entrepreneur in stagecoaching. The set employed imagery derived from Birch's business, California history, and California symbolism. The tea tray, for example, is engraved with pictures of Sutter's Mill, Sutter's Fort, and the San Francisco waterfront. The vegetable dish has a stagecoach for a finial and bears' heads as handles, and is engraved with a miner's tent set in front of distant mountains.[10] The new abundance of silver compelled the *nouveau* to display their *riches* in this grandiose fashion.

These large services had their counterpart in the small objects that began to be made in quantity; almost anything might now be fashioned in either sterling or plate. Trade catalogs of the time abound with small items; the ubiquitous souvenir spoons, like those made by Hammersmith & Field or Shreve, are good representatives of the expansion of silver into the more banal corners of the decorative arts.[11]

Despite these changes, some aspects of silver's role have remained essentially stable over the years. A good example is its incessant use for presentation pieces. Silver's durability and intrinsic worth, along with its capacity to take engraved inscriptions, have made it a preferred material for the purpose. They were particularly common in the second half of the last century, both in expensive and elaborate commissioned objects, and in what could be called a "vernacular" style of presentation piece, usually a

Detail, the Birch silver

Detail, the Birch silver

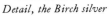

Souvenir spoons

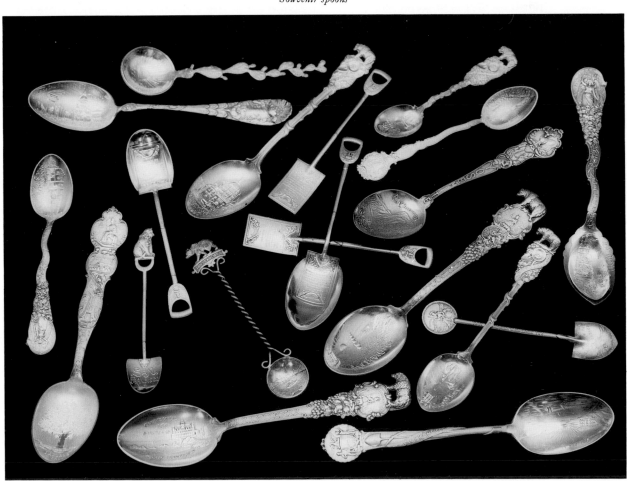

small beaker, cup, or vase of production-line quality individualized only by an engraved message. In 1858, Vanderslice and Company advertised that they were "prepared to furnish all Premiums for State Fairs and County Fairs" in such vernacular form.[12]

At their best, presentation pieces were awards bestowed to acknowledge heroism, commercial achievement, or some other accomplishment for the social good. In less altruistic cases, they may have served as a form of payment in kind, when remittance in specie would have seemed at best ill-mannered or at worst illegal. A gift of this type was intended to place certain obligations on the recipient to influence his behavior, preferably to reciprocate in a way agreeable to the giver.[13] This may perhaps have been the case with a large urn presented in 1860 to Edward Dickinson Baker "by the Merchants of San Francisco as a token of their esteem and confidence." Commissioned from a San Francisco jeweler and retailer, this presentation was designed (probably) to secure Baker's support for the construction of a transcontinental railroad and a bridge across the Golden Gate. A more typical presentation piece, and one less fraught with coercive overtones, was the commemorative bowl made by George C. Shreve and Company and presented in 1888 to Captain J. C. Hunter by grateful passengers aboard the ship *George W. Elder*.

An unusual form of presentation silver is naval vessel stateroom silver, intended for use by a ship's officers. The form is represented in the California experience by several services, including one made by Shreve for the cruiser *San Francisco*, and now aboard the aircraft carrier *Carl Vinson*. Ordered by the people of San Francisco, its decoration, including golden bears as finials, is characteristic of the topical motifs chosen by a city or state when they commissioned these services. Identical in purpose, yet attuned to a more modern era and vastly different in appearance, is the set made by Gorham for the USS *Long Beach*. Designed by Richard Huggins in 1961 on behalf of the citizens of the city of Long Beach, this set features "atomic models with depictions of electron paths" as its finials.[14]

Silver continues to be used for presentation pieces—witness the Lombardi Super Bowl trophy—but not with its former frequency, no longer playing the vibrant role in our lives it once did. It is too much trouble to keep clean, it is too easily stolen, there are too many substitutes. A further manifestation of the diminishing presence of silver in everyday life is the formation of collections of antique silver from various eras and locales. Displayed in museums as works of art and craft behind glass and barriers, these collections testify to the changing role of silver in the twentieth century.

California is particularly rich in collections. Mrs. John Emerson Marble, for example, began to collect early American silver in the 1920's with the help of the distinguished curator Gregor Norman-Wilcox; over the years she amassed a trove of more than four hundred objects. Since 1976 they have been in the care of Stanford University, Mrs. Marble's alma mater. (Coincidentally, Mrs. Marble graduated from Stanford in 1897, the same year the famous silver collector Francis Patrick Garvin received his degree from Yale.) A fine collection of English silver resides at the Huntington Library and Art Gallery in San Marino, and the Arthur and Rosalinde Gilbert collection of monumental English silver is housed at the Los Angeles County Museum of Art.[15]

The Baker silver

Captain Hunter's bowl

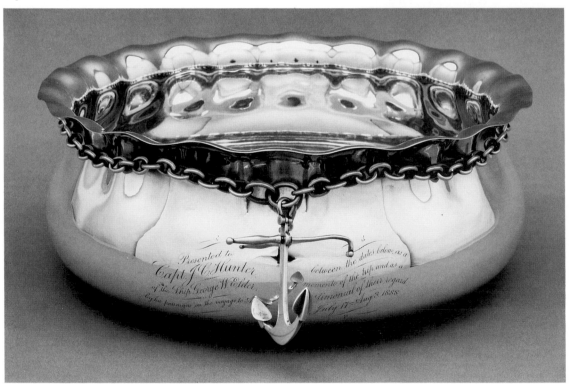

It remains to be seen whether silver's traditional role remains strong enough to sustain it as an icon of continuity in the future California and American experience. Deeply ingrained in our common speech—the glib have silver tongues, every cloud a silver lining, and the rich are born with silver spoons in their mouths—silver is still richly evocative. A recent study by two sociologists from the University of Chicago, Mihaly Csikszentmihalyi and Eugene Rochberg-Halton, may point towards an answer. Their study of the attitudes of eighty Chicago families towards their household goods concluded that "the[ir] most cherished objects" were those that evoked memories, associations, and experiences of the owner and his or her family and friends. The intrinsic qualities and stylistic attributes of an object were of secondary importance; the objects highly valued in psychic terms were those that provided links across generations, reminders of the ties that bind us to others. Csikszentmihalyi and Rochberg-Halton found that the photograph most frequently served this purpose for the modern family.[16] But using their analysis, we could also safely conclude that silver has been—and still can be—an important personal "icon of continuity." Engraved with the name of an ancestor or a family coat of arms, a silver object can be passed through time, proclaiming its history through successive generations. Like a photograph, a silver object can "provide a record of one's life, and of the lives of one's ancestors, and . . . be handed down to one's descendants."

It is likely this aspect of silver which has drawn Californians to the metal. Particularly in the nineteenth century, the outpouring of goods from California producers, as well as imports, probably belied more than just acquisitiveness. In a state mostly populated, even today, by immigrants, silver's traditional role as signifier of status and a sense of past could give its owner a kind of social legitimacy and personal continuity. By purchasing a tea set made of local silver by a local maker, or by commissioning a presentation piece replete with California images, the immigrant endowed his new land and society with some of the trappings of the old. Speaking its artifactual language of stability and heritage, silver lent its owner a sense of heritage and continuity in a state continually breaking new barriers and leading the rest of the country into the future.

X-Ray Fluorescence of California Silverware:

Some Preliminary Findings

JOHN W. BURKE

INTERESTING INFORMATION about a silver object may be gleaned from an analysis of its metallic composition. For example, traces of gold can help determine an object's age, since antique silverwares usually have higher gold contents than more recent ones, made from silver more efficiently refined. Yet attempts to determine *where* a certain piece was made, based on its elemental composition, have met with little success. Most old European and East Coast silverware was fabricated from melted coins, or remade from out-of-style or damaged wares. These varied sources, along with diverse alloying formulas, have generally made impractical the localization of an object's site of manufacture.

After the 1860s, the majority of silver used by California silversmiths originated in the West, both from the great silver and gold-quartz deposits of the Comstock, and from the Cerro Gordo mines in southeastern California. The Cerro Gordo mines, in particular, were the prime supplier to Thomas Selby, the major source of silver bullion for San Francisco silversmiths from 1869 until the mid-1870s. Selby bought Cerro Gordo lead ores for his flourishing foundry business, but also crudely extracted the silver in it (present in the ore at less than three per cent of its weight) at his refinery on Hyde Street in San Francisco. Although silver was only a by-product, his lead smelting volume was such that Selby was able to underprice his competition and dominate the local market for silver bullion.

The refining processes for lead and gold ores differ substantially. Aware of this, and recalling Selby's near-monopoly, Edgar W. Morse, consulting curator to the *Silver in the Golden State* project at the Oakland Museum, suggested that there be an investigation into the elemental composition of nineteenth century San Francisco silverware. It was hoped that such an inquiry might yield data useful in the attribution of the works of California silversmiths of the period. It has long been known that nineteenth century silversmiths, both in California and elsewhere, marked not only their own works, but also pieces mass-produced in East Coast factories, which the smith then resold. Any distinguishable difference between the composition of eastern and western silver might

99

therefore help determine if a California smith actually made a particular piece bearing his mark.

The author used energy dispersive X-ray fluorescence (XRF) to analyze eighteen pieces of California-made flatware, c. 1855 through c. 1880. (See Table 1.) Flatware was chosen because design characteristics peculiar to a given silversmith can reliably establish him as its maker. The XRF method is extremely sensitive, and non-destructive of the sample. Exposure to weak X-radiation causes the different elements within a tested object to fluoresce at characteristic frequencies and at amplitudes proportional to the quantities of the elements present. Although XRF has been used (principally at the Winterthur Museum Analytical Laboratory) to analyze thousands of eastern silver objects, the author could find no reports of tests on West Coast silverwares.

Although it was not anticipated that the percentages of total silver in the tested pieces would corollate with manufacturing location, those percentages were determined in the course of analysis. The results (see Table 2) confirmed the prediction; silver content varies widely from piece to piece, and cannot therefore reliably indicate the site of a piece's origin. What is interesting about this particular analysis, however, is the light it sheds on quality-control practices of the silversmiths represented in the sample. Presumably, objects marked "coin" should be at least ninety percent silver, and those marked "sterling" should meet or exceed the British silversmiths' guild standard of ninety-two-and-a-half percent silver. Unconstrained by the British standard, however, it was not uncommon for nineteenth century American silversmiths to misrepresent the quality of their wares. Our analysis revealed that *none* of the five pieces bearing quality stamps were up to their asserted content; in particular, a Vanderslice "Comstock" teaspoon marked "sterling" was less than seventy-five percent silver!

The inconclusiveness of total silver content notwithstanding, it *was* hoped that an examination of trace elements in the pieces might furnish an index of origin, a distinguishable "fingerprint." The result of trace element analysis of the California objects was compared with the same results for seventy pieces of eastern manufacture. Figures 1 and 2 illustrate the relative amounts of gold and lead, expressed as percentages of silver content, for both the eastern and western samples. Taken alone, these values are not distinctive for either group; the lead content of the California pieces appears marginally higher, and the gold lower, than in eastern pieces. It may also be observed that the gold content of the eastern samples decreases over time, and that the western pieces fall but slightly below this trend.

When the *ratio* of relative gold and lead contents is taken, however, a difference between the two groups does appear to emerge; the values—although there is a broad variation within the sample group—are appreciably greater for the California silver. The average lead-to-gold ratio is approximately six times greater for the California pieces than for the eastern; this is illustrated in Figure 3.

It should be emphasized that these results are preliminary. The sample of western silver is too small (eighteen pieces) and too restricted with respect to period (c. 1855 through c. 1880) to allow firm conclusions to be drawn. Furthermore, not only do nearly all the eastern pieces used for comparison predate the California objects, but we also cannot be certain whether or not the tested western pieces were made from coins or melted-down eastern silver. Clearly, many more western objects must be analyzed before generalizations may safely be made (particularly, the author suggests that an examination of other trace elements [e.g., arsenic or bismuth] might yield content patterns which corollate closely with site of origin).

Nevertheless, based on the limited evidence at hand, there does appear to be a difference, in the relative amounts of trace lead and gold, between silver of western and eastern origin. Assuming that further research confirms this as characteristic of site of origin—whether attributable to the source of the ore or to the method of refining—such a ratio may prove useful in corroborating attribution of silverware to western makers.

Table 1:

Artifacts Analysed

	Manufacturer	Pattern	Date
a.	Eaves & Nye	"Cincinnati" Tbs.	1850s—1860s
b.	Eaves & Nye	"Cincinnati" tsp.	1850s—1860s
c.	Lawler	"Plain" tsp.	c. 1860s
d.	Koehler & Ritter	"Eugene" tsp.	c. 1860s
e.	Vanderslice	"Bead" tsp.	c. 1860s
f.	Reichel	"Eugene" tsp.	before 1868
g.	Koehler & Ritter	"Medallion" tsp.	c. 1860s
h.	Schulz & Fischer	"Antique" Tbs. *	early 1870s
i.	Reichel (K&R)	"Gothic" tsp.	early 1870s
j.	Vanderslice	"Gargoyle" Tbs.	before 1875
k.	Vanderslice	"Comstock" tsp. **	before 1875
l.	Vanderslice	"Gargoyle" tsp.	before 1875
m.	Koehler & Ritter	"Marin" Tbs.	c. 1870s
n.	Lawler	"Plain" tsp.	c. 1870s
o.	Schulz & Fischer	"Medallion" tsp. **	late 1870s
p.	Schulz & Fischer	"Faralone" tsp. **	late 1870s
q.	Vanderslice	"Comstock Grap." tsp.	after 1875
r.	Schulz & Fischer	"5 star" tsp. **	c. 1880

*Stamped "Coin"   **Stamped "Sterling"

Table 2:

Results of XRF Analysis

Amounts expressed as percent, T=trace

	Silver	Copper	Gold	Lead	Iron	Nickle	Zinc
a.	88.54	10.02	0.08	0.35	0.01	0	0.25
b.	89.06	9.59	0.09	0.25	0.01	0	0
c.	83.06	15.03	0.39	0.5	0.02	T	0
d.	90.01	8.45	0.05	0.47	0.02	0	0
e.	90.54	8.1	0.03	0.31	0.02	0	T
f.	87.63	10.49	0.11	0.74	0.03	0	T
g.	74.52	23.49	0.12	0.82	0.05	T	0
h.	88.6	9.65	0.02	0.72	0.01	0	T
i.	88.03	10.14	0.09	0.51	0.02	0	0.2
j.	89.85	8.65	0.01	0.46	0.03	0	0
k.	74.28	23.78	0.08	0.81	0.05	0	0
l.	90.6	7.77	0.02	0.34	0.02	0	0
m.	89.36	9.38	0.01	0.24	0.02	0	0
n.	91	7.62	0	0.35	0.04	T	T
o.	90.55	8.26	0.01	0.16	0.02	0	0
p.	89.98	8.67	0.04	0.3	0.02	T	T
q.	91.64	6.45	0.05	0.85	0.01	0	0
r.	88.12	10.27	0.07	0.53	0.02	T	0

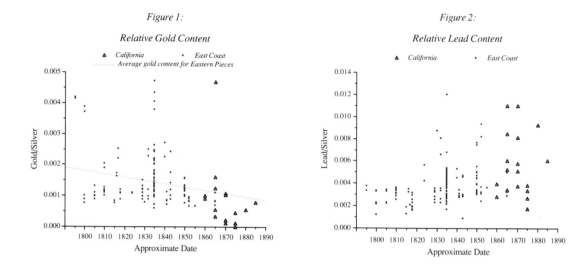

Figure 1:

Relative Gold Content

Figure 2:

Relative Lead Content

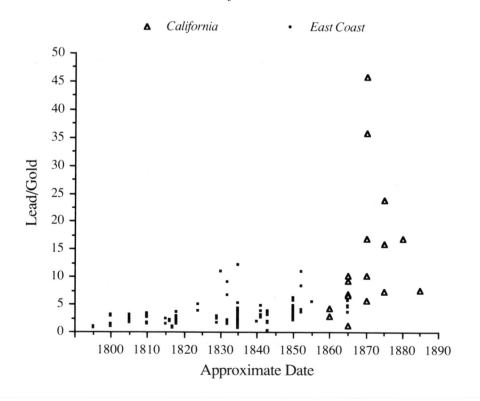

Figure 3:

Ratio of Lead to Gold

△ California   ▪ East Coast

Northern California Flatware Patterns

EDGAR W. MORSE

DRAWINGS BY BARRY HOPP

THIS ILLUSTRATED LIST includes all the Northern California flatware patterns that have been identified.[1] No catalogs for the major nineteenth-century makers exist; many Schulz & Fischer patterns are illustrated in a leaflet they published c. 1878.[2] Two other Schulz & Fischer patterns are identified in a Hammersmith & Field catalog from the 1880s.[3] Many of Shreve's twentieth-century patterns are identified in their four catalogs and in leaflets published c. 1946–60. Collectors and dealers have assigned names to patterns where documentary evidence is absent. In a few cases, alternative names have been used; the less-common name follows a slash ("/").

In addition to their unique designs, San Francisco silversmiths produced flatware in patterns that were standard throughout the United States. These included "Plain" / "Fiddle," "Plain Tipped," "Threaded" / "Plain Threaded," "Antique," "Windsor," and "Beaded."[4] In the absence of other evidence, it is usually difficult to decide who made the standard patterns, for the marks are deceptive (c.f. APPENDIX II).

Elliot Evans and Paul Evans have devoted considerable efforts to date the introduction of different patterns; these dates are given in brackets, and should be considered approximate only. In the case of several of the later Schulz & Fischer patterns, documentary evidence exists; the dates for these are given without brackets.

(In the following list of makers and their patterns, illustrations are furnished for those patterns that are unique to a maker.)

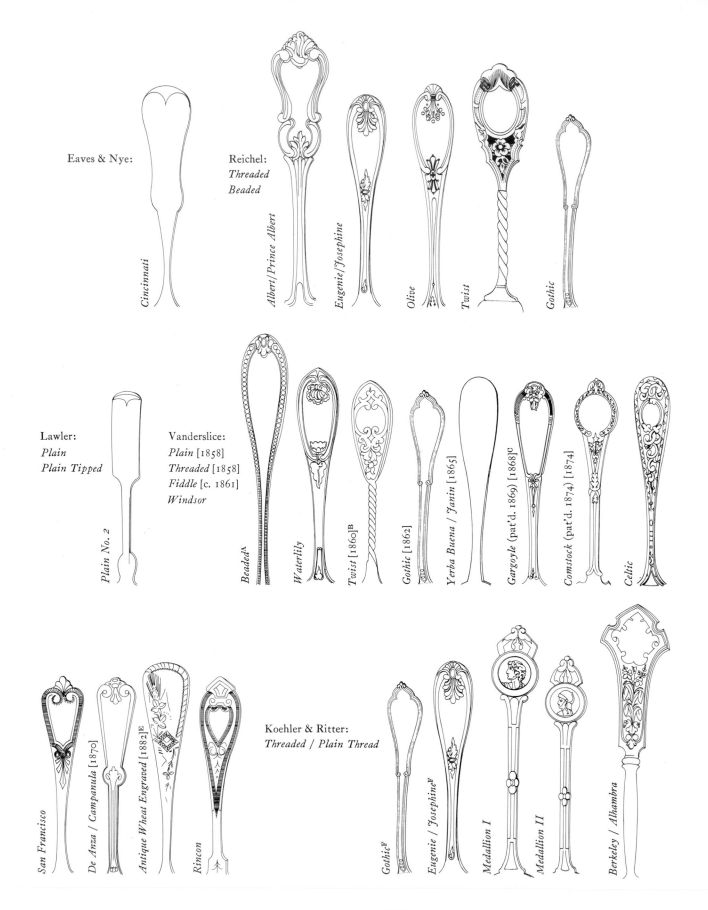

Eaves & Nye:

Cincinnati

Reichel:
Threaded
Beaded

Albert / Prince Albert

Eugenie / Josephine

Olive

Twist

Gothic

Lawler:
Plain
Plain Tipped

Plain No. 2

Vanderslice:
Plain [1858]
Threaded [1858]
Fiddle [c. 1861]
Windsor

Beaded[A]

Waterlily

Twist [1860][B]

Gothic [1862]

Yerba Buena / Janin [1865]

Gargoyle (pat'd. 1869) [1868][C]

Comstock (pat'd. 1874) [1874]

Celtic

San Francisco

De Anza / Campanula [1870]

Antique Wheat Engraved [1882][E]

Rincon

Koehler & Ritter:
Threaded / Plain Thread

Gothic[F]

Eugenie / Josephine[F]

Medallion I

Medallion II

Berkeley / Alhambra

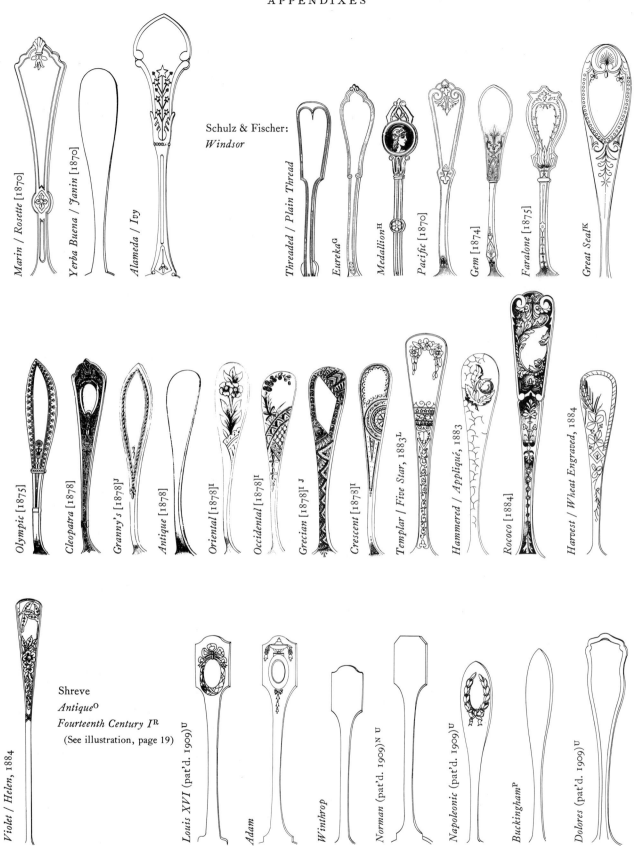

Marin / Rosette [1870]

Yerba Buena / Janin [1870]

Alameda / Ivy

Schulz & Fischer:
Windsor

Threaded / Plain Thread

Eureka[G]

Medallion[H]

Pacific [1870]

Gem [1874]

Faralone [1875]

Great Seal[K]

Olympic [1875]

Cleopatra [1878]

Granny's [1878][J]

Antique [1878]

Oriental [1878][I]

Occidental [1878][I]

Grecian [1878][I] [J]

Crescent [1878][I]

Templar / Five Star, 1883[L]

Hammered / Appliqué, 1883

Rococo [1884]

Harvest / Wheat Engraved, 1884

Violet / Helen, 1884

Shreve
Antique[O]
Fourteenth Century I[R]
(See illustration, page 19)

Louis XVI (pat'd. 1909)[U]

Adam

Winthrop

Norman (pat'd. 1909)[N] [U]

Napoleonic (pat'd. 1909)[U]

Buckingham[P]

Dolores (pat'd. 1909)[U]

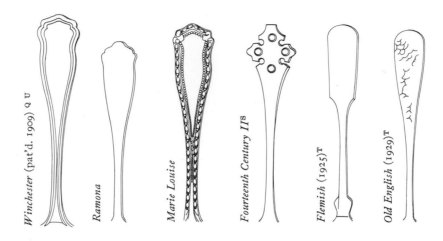

Winchester (pat'd. 1909) Q U   *Ramona*   *Marie Louise*   *Fourteenth Century II* S   *Flemish (1925)* T   *Old English (1929)* T

1   This Appendix is based on, and brings up to date, articles by the author: "Schulz & Fischer: San Francisco Silversmiths" (San Francisco, 1977), and " 'San Francisco' Victorian Flatware Patterns: Vanderslice, Koehler & Ritter, F. R. Reichel, and Others" (San Francisco, 1977). The author is also indebted to articles appearing in *Spinning Wheel* by Elliot Evans and Paul Evans: "San Francisco Silver: Schulz & Fischer Flatware" (May, 1977), pp. 19–22; "San Francisco Silver: Vanderslice & Co. Flatware" (December, 1977), pp. 18–20; "San Francisco Silver: F. R. Reichel and Koehler & Ritter Flatware" (March, 1978), pp. 36–38; and "Shreve & Co. of San Francisco Flatware" (September, 1978), pp. 30–32. This Appendix is also based on examples in the collections of Argentum Antiques, Ltd., of San Francisco, and in the Oakland Museum History Department.

2   A copy of the leaflet is in the collections of the Bancroft Library at the University of California, Berkeley.

3   Collections of Gary Breitweiser, Santa Barbara, California.

4   Noel Turner, *American Silver Flatware 1837–1910* (San Diego, 1972), p. 85.

CODES: A—Vanderslice made two different versions of *Beaded*; the second dates from about 1870. B—Vanderslice made many versions of *Twist* that vary in engraving and the shape of their tops. This appears not to have been a line pattern. C—Design patent no. 3,402. Pieces exist "Pat. Ap'd For." D—Design patent no 7,776. E—This is the most common pattern including *Antique Poppy Engraved* and *Antique Rose Engraved*. Unengraved versions exist with or without a reverse tip and satin finish. F—These appear to have been made using Reichel's dies. G—Almost identical to the *Gothic* of Reichel and Koehler & Ritter. H—Misattributed in Turner, *American Silver Flatware*, 119[4]. I—Uses the *Antique* blank. J—No examples are recorded. K—This is the only Schulz & Fischer flatware pattern that is marked with their mark for holloware. (C.f. APPENDIX II.) L—Hammersmith & Field Catalog. M—John G. Fox collection; invoices; Nevada Historical Society, Reno. N—Also supplied with hammered surface, beveled edges, and applied initials. O—Also supplied as *Antique* (plain), *Antique* (hammered), and *Windsor* (with engraved thread). P—Supplied plain or with applied initial. Q—Original name was *Italian*. R—Offered with twelve different cut-out handles to each dozen. This pattern appeared only in the 1911–12 catalog. S—Replaced the earlier version with the same name in the 1912–13 catalog. T—Designed by Porter Blanchard. U—Designed by Joseph E. Birmingham.

Marks on California Silver

EDGAR W. MORSE

DRAWINGS BY BARRY HOPP

THIS IS A LIST of the most common marks to appear on California silverware. It makes no pretense to be complete, and many retailers' marks have been deliberately omitted. During the nineteenth century, California silversmiths, in common with the practice throughout the country, supplied goods either with their own marks, theirs plus a retailer's name, the retailer's name only, or no marks at all. California silversmiths at times also put their marks on pieces marked by eastern manufacturers. As a result, it is not possible on the basis of marks alone to be completely certain about who made any particular piece. However, the California silversmiths *did* make large quantities of silverware. As in the case of paintings, with knowledge and effort attributions can usually be made with reasonable confidence. In the list that follows, changes in marks are useful for dating; the business code attempts to separate actual silversmiths from those who retailed the work of others under their own names; locations are the principal locations; and dates are often approximate for want of documentation.

NAME, MARK AND CODE:	LOCATION:	DATE:
Allan Adler, 1	Los Angeles	c. 1938–present
Anderson & Randolph, 4	San Francisco	1871–1877
Barrett & Sherwood, 4	San Francisco	1849–post-1902
John O. Bellis, 1, 3	San Francisco	1907–1943
William Benbow & Co., 1	Los Angeles	1932–post-1940
Porter Blanchard, 1:	Burbank, Los Angeles, Calabasas	1923–present

Braverman & Levy, 4	San Francisco	c. 1856–1881
Harry S. Dixon, 5	San Francisco	c. 1919–post-1940
Alf/red/H. Ellsworth, 1	Pasadena	1925–1929

John Field, 1	Santa Barbara	1901–c. 1930
Clemens Friedell, 1	Pasadena	1912–1920; 1928–1963
Hammersmith & Field, 1, 4 /	San Francisco	1886–1906
Hammersmith & Co., 4	San Francisco	1907–1912
Haynes & Lawton, 9	San Francisco	c. 1869–c. 1873
Henry Hoppe, 1	San Francisco	c. 1910–c. 1941
Jacks & Woodruff, 1, 4/	San Francisco	1849–c. 1853
Jacks & Brothers, 1, 4		
Joseph Brothers, 4, 6	San Francisco	1852–c. 1859
Koehler & Ritter, 1, 4:	San Francisco	1867–1884

George Ladd, 2	San Francisco	1856–1861
William Lawler, 1, 2, 7	San Francisco	c. 1853–1881
M & M Silversmiths, 1	Los Angeles	1924–post-1940
Myllo Brothers, 1	Los Angeles	1921–1939
Frederick R. Reichel, 1, 3	San Francisco	c. 1856–1867
Robert Schaezlein/	San Francisco	1882–post-1940
Schaezlein & Burridge, 1		
Schulz & Fischer, 1:	San Francisco	1867–1890
Fischer & Schulz		1867–1868
Schulz, Fischer & Mohrig		1869–1873
Schulz, Fischer & McCartney		1883–1887
Schulz & McCartney		1888–1890

Karl Sedlacek, 1	Los Angeles	1918–post-1940
G. C. & S. S. Shreve, 2	San Francisco	1852–1857
George C. Shreve & Co., 4	San Francisco	1858–1883
George C. Shreve & Co., 1, 4	San Francisco	1883–1893
Shreve & Co., 1, 4	San Francisco	1894–1968

1883–1909

1909–1922

Shreve & Co., 2	San Francisco	1968–present
Shreve, Treat & Eacret, 2	San Francisco	c. 1914–1941
Arthur Thumler, 1/	San Francisco	1920–1969
Thumler Brothers, 1, 2, 10:		

J. W. Tucker & Co./	San Francisco	1850–c. 1885
Tucker Jewelry & Mfg. Co., 2, 7		
Peter Traphagen, 5	Riverside, Pasadena, etc.	c. 1934–post-1940
W. K. Vanderslice & Co., 1, 2:	San Francisco	1858–1908
	134 Washington St.	1858–1860
	728 Montgomery St.	1860–1862
	810 Montgomery St.	1863–1871
	136 Sutter St.	1872–1906

Dirk Van Erp, 5, 8	San Francisco	1910–1933
William Van Erp, 1, 5, 8	San Francisco	1933–1977
(Used the name "Dirk Van Erp")		

BUSINESS CODES: 1—Silversmith   2—Retail jeweler   3—Manufacturing jeweler   4—Retail and manufacturing jeweler   5—Art metal worker   6—Wholesale jeweler   7—Watchmaker 8—Coppersmith   9—Agents for the Pacific Plate Works, and dealers in crockery   10—Engravers

Notes

I am much indebted to Michael J. Weller for his assistance throughout the research of this article, and especially for editing this manuscript. Biographical data is mostly from city directories, the California *Great Registers*, or the Federal or California Censuses of Population. I refer to standard references for the Eastern careers of various jewelers and silversmiths.

1 L. Anderson, *Art of the Silversmith in Mexico* (New York, 1975), pp. 4–7.

2 G. W. James, *In and Out of the Old Missions* (Boston, 1927), pp. 81, 249, 265. K. Baer, *Treasures of Mission Santa Inés* (Fresno, 1956), pp. 277, 278.

3 San Francisco *Alta California* (Aug. 20, 1850), p. 4; D. Webster, *Writings and Speeches* (1903), p. 361. R. G. Dun & Co. "Credit Ledgers," "California" *A*, 5. These ledgers are at Harvard University, Manuscripts Department, Baker Business Library. San Francisco *Examiner* (May 30, 1890), p. 2.

4 "John W. Tucker" in O. T. Shuck, *Leading and Representative Men* (San Francisco, 1875), pp. 709–711; C. Carpenter, *Gorham Silver* (New York, 1982), pp. 61–62.

5 San Francisco *Bulletin* (December 16, 1858), p. 2; (Jan. 21, 1861), p. 2; San Francisco *Call* (July 28, 1899), p. 1; *Jewelers' Circular* (Aug. 2, 1899), p. 12.

6 *Jewelers' Circular* (Mar. 29, 1893), (May, 1885), p. 105; F. H. Hackett, *Industries of San Francisco* (San Francisco, 1884), p. 115; *The Argonaut* (Feb. 15, 1881), p. 16; *Bulletin* (Nov. 2, 1885), p. 2; *Jewelers' Circular* (Jan. 1891), p. 84.

7 *Report of the First Industrial Exhibition of the Mechanics' Institute* (San Francisco, 1857), p. 91; San Francisco *Chronicle* (Dec. 19, 1867), p. 4; 1870 Census of Industry, ms Schedule *4*, San Francisco Ward 3, p. 22. These schedules are at the California State Library in Sacramento.

8 *Bay of San Francisco* (Chicago, 1892), *2*, p. 666; S. Vanderslice, *Vanderslice and Allied Families* (Los Angeles, 1931), pp. 38–39, 47–48; San Francisco *Chronicle* (April 20, 1954), p. 29.

9 *Report of the Second Industrial Exhibition . . .* (San Francisco, 1858), pp. 19, 29, 88; *Transactions* of the California State Agricultural Society (1859), 247 (1873), pp. 12, 14, 23; (1891), p. 16; *Victoria Directory*, "San Francisco Department," pp. 6, 30; *Alta* (Jan. 21, 1865), p. 1; *ibid.* (Dec. 11, 1868), p. 1. Property figures from 1870 Census; R. G. Dun, "California," *14*, p. 5.

10 *Alta* (Dec. 12, 1868), p. 1; Dun, "California," *14*, p. 5 (Nov., 1869), 1870 Census of Industry *3*, Schedule 4, np; Dun *loc. cit.* (Aug. 25, 1871), (Sep. 3, 1873), (Sep. 12, 1874), (April 22, 1875), (Dec. 8, 1869), (Dec. 8, 1876); *Alta* (Jan. 1, 1875), p. 2; I have found

between nine and fourteen of his employees in the directories for the 1870s. This relative constancy indicates a stable business.

11 G. H. Smith, *History of the Comstock Lode* (Reno, 1942), pp. 188–194, 207; E. Lord, *Comstock Mining and Miners* (Berkeley, 1959), p. 416.

12 California Supreme Court #9541, *Daggett* v. *Vanderslice*, Filed April 23, 1884, p. 46; *Bulletin* (Aug. 16, 1880), p. 2; 1880 Census of Industry, San Francisco Schedule 3, "Disturnell," p. 12; *Bulletin* (Aug. 16, 1880), p. 2; California Supreme Court #11535, *Vanderslice* v. *Traylor* filed March 27, 1876, pp. 34–35; *Bulletin* (Jan. 1, 1881), p. 2; Melrose first worked for Vanderslice as an engraver in 1875. He was born about 1841 in England.

13 Daniel Murden started as an apprentice in 1887. He started working for Shreve as a silversmith by 1892; *Argonaut* (Aug. 25, 1883), p. 13. The copy uses English spelling, which suggests that Melrose wrote it; *ibid.* (Dec. 20, 1884), p. 19; (Nov. 28, 1885), p. 11; *Jewelers' Circular* (May 17, 1893), p. 15; *ibid.* (Dec. 4, 1895), p. 9; *Call* (Mar. 13, 1899), p. 9.

14 *Jewelers' Circular* (March 11, 1896), p. 31.

15 *Report of the Committee of Five to the "Thirty-Five Companies" on the San Francisco Conflagration* (New York, 1906), p. 59; *Call* (Nov. 28, 1908), p. 7; *Jewelers' Circular* (Jan. 27, 1909), p. 100.

16 *Jewelers' Circular* (Nov. 16, 1898), p. 1.

17 *Bulletin* (Nov. 19, 1867), p. 3; *San Francisco Municipal Reports 1867–68*, "Public Administrator's Report," p. 381.

18 R. G. Dun, "California," *14*, p. 355 and *19*, p. 5; 1870 Census of Industry, Schedule 3, San Francisco Ward 3, p. 25.

19 R. G. Dun, *loc. cit.*; *Mining and Scientific Press* (May 1, 1875), p. 293; *Report of the Tenth Industrial Exhibition* (San Francisco, 1875), p. 161; *Report of the Eleventh Industrial Exhibition* (San Francisco, 1876), p. 126.

20 *Bulletin* (Aug. 24, 1878), p. 2; *Jewelers' Circular* (March, 1879), p. 30; (April, 1879), p. 4. Internal evidence indicates that the anonymous creditor was Levison Brothers; 1880 Census of Industry, San Francisco, Schedule 3, "H. G. Langley," p. 19; *Jewelers' Circular* (Jan. 16, 1895), p. 35.

21 R. G. Dun, "California," *14*, p. 6; 1870 Census of Industry, Schedule 4, San Francisco, First precinct, 7th Ward, "W. H. Brown," p. 3.

22 R. W. Hansen, "The Hewes Receipt," *Keepsake Number 12* (San Francisco: Book Club of California, 1969). Contrary to what Hansen assumes, Mohrig was not active in the business then or ever; E. Sabin, *Building the Pacific Railway* (1919), p. 208; R. G. Dun, "California," *16*, p. 377; Sotheby's–New York (Nov. 19, 1983), *3*, Lot 525. John G. Fox Collection, Nevada Historical Society, Reno, bound book of invoices. An invoice dated May 26, 1875 is the last to say "Solid Silverware"; one dated June 30, 1875, is the first to say "Sterling." *Report of the Tenth Industrial Exhibition* (San Francisco, 1875, p. 161; *Mining and Scientific Press* (Sept. 25, 1875), p. 197; R. G. Dun *16*, p. 377 (Feb. 8, 1876).

23 Internal evidence in this leaflet at the Bancroft Library yields the 1878 date; the range of pieces in one pattern, "Cleopatra," is indicated in E. Evans, " 'Cleopatra' Sterling of Schulz & Fischer," *Silver 7* (Jan.–Feb., 1974), pp. 14–17. I am indebted to Charles Fuller for letting me see this set and for much assistance. The "working ends" of their other patterns are usually identical to those of "Cleopatra."

24 J. S. Hittell, *Commerce and Industries of the Pacific Coast* (San Francisco, 1882), pp. 696, 698; California Bureau of Labor Statistics, *Second Biennial Report for 1885 and 1886*,

p. 615. McCartney lived for a time at Fischer's address and married one of Fischer's daughters; *Alta* (Jan. 6, 1890), p. 7.

25 G. S. Gibb, *Whitesmiths of Taunton* (New York, 1969), pp. 230–231; United States Interstate Commerce Commission, *Railways in the United States in 1902: Part 2* (Washington, 1903), p. 154.

26 *Jewelers' Circular* (Oct. 23, 1893), p. 13; (Aug. 26, 1896), p. 15; *Alta* (Aug. 15, 1853), p. 1; R. G. Dun, "California," *16*, p. 374.

27 *Bulletin* (Feb. 5, 1881), p. 1; (Mar. 22, 1881), p. 1; *Jewelers' Circular* (Oct., 1882), p. 306; *Bulletin* (Feb. 10, 1883), p. 3; (Sept. 15, 1883), p. 3; *Chronicle* (Aug. 19, 1883), "Quadruple Sheet," p. 4. Collectors of San Francisco silver have long debated the significance of the "Bee" mark. Its origins remain obscure, but it was used as Shreve's trademark from 1883. It is quite unlikely that it was originated by Frederick Witt, for he worked for Schulz & Fischer from 1871 to 1885, and was then self-employed until he became a silversmith for Shreve in 1888.

28 *Bulletin* (Oct. 8, 1885), p. 3; (Oct. 9, 1885), p. 2; (Oct. 12, 1885), p. 2; San Francisco Directory for 1886, "Advertising Section," p. 37; *Jewelers' Circular* (Oct. 25, 1893), p. 13 (Nov. 1, 1893), p. 23.

29 *Ibid.* (Feb. 7, 1894), p. 133; California Historical Society, Shreve and Company file, manuscript, "Reminiscences of James Daniel Murphy, aged 12 in 1898"; *Jewelers' Circular* (April 28, 1897), p. 24; (June 15, 1898), p. 31; (Aug. 23, 1899), p. 39; (Nov. 15, 1899), p. 42; (Dec. 6, 1899), p. 38; (Sep. 17, 1902), p. 72; *Report of the Committee of Five . . .*, pp. 34–35.

30 *Sunset* (April, 1907), pp. 530–531; *Call* (Oct. 16, 1906), p. 9; *Jewelers' Circular* (May 13, 1908), p. 73; (Aug. 26, 1908), p. 79.

31 They also published catalogs for 1910–1911 and 1912–1913; *Jewelers' Circular* (Oct. 21, 1908), p. 101; *Examiner* (April 15, 1913), p. 27.

32 *Chronicle* (June 28, 1921), p. 10; E. Evans, "Shreve & Co. of San Francisco Flatware," *Spinning Wheel* (Sep., 1978), pp. 30–32; Noel Turner, *American Silver Flatware 1837–1910* (San Diego, 1972), p. 85.

33 *Facets* (June–July, 1978), p. 2: this was a house organ of Shreve's then owner, Dayton Hudson.

34 *Jewelers' Circular* (July 22, 1908), p. 71.

35 This account of Arthur Thumler relies heavily on information in Fred H. Koster, Jr. "Thumler Brothers—San Francisco Silversmiths," an undated and unpublished manuscript. I owe a great debt to the late Mr. Koster, who introduced me to the work of the Thumler Brothers and gave me a copy of his article. Other information from members of the family; *Chronicle* (April 8, 1966), p. 40; (Feb. 27, 1969), p. 35; (Aug. 24, 1980), p. 37.

LESLIE GREENE BOWMAN

1 Unfortunately, objects and information are scarce for William Benbow, Charlotte Crane, Alf Ellsworth, Eric Magnussen, Phillip Paval, Hudson Roysher, Karl Sedlacek, and Peter Traphagen; further research is needed. Some, like Allan Adler, Harry Osaki, and Lewis Wise began their careers at the end of the period covered by this book, and should be treated in a study of post-war silver.

NOTES

2 Born Clemens Friedl, he altered the spelling of his Germanic-sounding name during World
   War I to avoid persecution.

3 Charles Carpenter's research has shown that Martelé pieces were raised by one smith and
   chased by another, the latter paid a wage 12½ per cent more than the former. For more
   information on Martelé, see Charles H. Carpenter, Jr., *Gorham Silver, 1831–1981* (New
   York, Dodd, Mead & Co., 1982), pp. 221–252.

4 Friedell family papers.

5 "Comes Here for $15,000 Present," unrecorded citation from Pasadena newspaper, 1912.
   Friedell family papers.

6 *Ibid.*

7 *Pasadena Star*, Tournament of Roses Number, 1914.

8 Leonard Kreidt, "A Life Lined with Silver; Craftsman in Generous Gesture to City He
   Loves," *Pasadena Star News*, February 12, 1960.

9 *Pasadena Star*, Tournament of Roses Number, 1914.

10 *Maryland Huntington Life*, Bungalow Number, 1917, pp. 4–5.

11 Leonard Kreidt, "A Life Lined with Silver."

12 Porter Blanchard, "An Easterner Comes West," *California Southland*, 6:53 (May 1924),
   p. 24.

13 Private collection.

14 "New Art Shop" news clipping c. 1925, unrecorded citation, Blanchard family papers.

15 Porter Blanchard, "History of Silvercraft Reviewed by Artist Porter Blanchard," news
   clipping, unrecorded citation, Porter Blanchard Papers, Archives of American Art,
   Smithsonian Institution.

16 Marion Hugus Clark, "The Arts and Crafts Society," *California Southland*, 7:61 (January,
   1925), p. 11.

17 "M. U. S.," "The Arts and Crafts Society of Southern California," *California Southland*,
   6:56 (August, 1924), p. 24.

18 Porter Blanchard, "History of Silvercraft."

19 Marion Hugus Clark, "The Arts and Crafts Society," p. 1.

20 "Silversmith Plies Old Art," *Citizen-News*, c. 1925, Porter Blanchard Papers.

21 "Porter Blanchard Shop Displays Silver Craft," news clipping, unrecorded citation,
   Porter Blanchard Papers.

22 "Screen Stars Flock to Silversmith," clipping, unrecorded citation, Porter Blanchard
   Papers.

23 Undated manuscript, Porter Blanchard Papers.
24 Leonard Kreidt, "A Life Lined with Silver."

DONALD L. HARDESTY

1 An excellent history of the Comstock may be found in Eliot Lord, *Comstock Mining and
   Miners* (reprint of 1881 edition, Berkeley, California, Howell-North, 1959).

2 Harold Bonham, "Geology and Mineral Deposits of Washoe and Storey Counties,

Nevada," *Nevada Bureau of Mines and Geology Bulletin 70* (1969); Vincent Gianella, "Geology of the Silver City District and the Southern Portion of the Comstock Lode, Nevada," *Nevada Bureau of Mines and Geology Bulletin 29* (1936).

3   G. F. Becker, "Geology of the Comstock Lode and the Washoe District," *U.S. Geological Survey, Monograph Number 13* (1882).

4   Otis E. Young, Jr., "Philipp Deidesheimer, 1832–1916: Engineer of the Comstock," *Historical Society of Southern California 57*, pp. 361–369 (Winter, 1975).

5   Dan DeQuille (William Wright), *The Big Bonanza* (reprint of the 1876 edition, Las Vegas, Nevada Publications, 1974), p. 250.

6   Rossiter W. Raymond, *Statistics of Mines and Mining in the States and Territories West of the Rocky Mountains*, Forty-first Congress, Second Session, ex. doc. no. 207 (1870), pp. 55, 489–96.

7   *Ibid.*, Forty-second Congress, Third Session, ex. doc. no. 210 (1873), p. 119.

8   Works Progress Administration Nevada Writers' Project, manuscripts for each of the fifty-six Comstock mines summarizing contemporary newspaper accounts, *The Consolidated Virginia Mine*, p. 2, and *The Gould and Curry Mine*, p. 3 (1940–41).

9   Raymond, *Statistics*, 1873, p. 119.

10   P. 227.

11   Lord, *Comstock Mining*, p. 347.

12   *Ibid.*, p. 348.

13   DeQuille, *Big Bonanza*, p. 228.

14   Lord cites the *Territorial Enterprise*, September 13, 1863, for this information; no extant copies of that issue remain, so verification is not now possible.

15   DeQuille, *Big Bonanza*, p. 229.

16   *Ibid.*, p. 394.

17   *Ibid.*, p. 385.

18   Grant H. Smith, "The History of the Comstock Lode, 1850–1920," *Nevada Bureau of Mines and Geology, Geology and Mining Series 37* (Reno, University of Nevada, 1943), p. 46.

19   Lord, *Comstock Mining*, p. 393.

20   P. 230.

21   Lord, *Comstock Mining*, p. 89.

22   P. 231.

23   WPA, *Ophir Mine*, manuscript, p. 1.

24   Young, "Deidesheimer," p. 374.

25   Lord, *Comstock Mining*, p. 345; Smith, "History of the Comstock," p. 280.

26   DeQuille, *Big Bonanza*, p. 88.

27   John D. Galloway, "Early Engineering Works Contributory to the Comstock," *Nevada Bureau of Mines and Geology, Geology and Mining Series 45* (Reno, University of Nevada, 1947), p. 65.

28   *Ibid.*, p. 67.

29   Thompson and West, *History of Nevada* (1881), p. 68.

30 December 4, 1861.

31 Lord, *Comstock Mining*, pp. 124–125.

32 WPA, *Ophir Mine*, manuscript, p. 2.

33 Lord, *Comstock Mining*, Chapter II, pp. 15–32, has a good account of the early miners in Gold Canyon.

34 DeQuille, *Big Bonanza*, pp. 296–297.

35 Lord, *Comstock Mining*, pp. 368–369.

36 P. 198.

37 Lord, *Comstock Mining*, p. 383.

38 *Ibid.*, p. 373.

39 Paul Fatout, *Mark Twain in Virginia City* (Bloomington, Indiana University, 1964), p. 69.

40 DeQuille, *Big Bonanza*, p. 248.

41 Lord cites a wage of $4.75 a day for deep work, as opposed to the normal wage of slightly over $4.05 a day (*Comstock Mining*, p. 387).

42 *Ibid.*, p. 396.

43 *Ibid.*, p. 185.

44 Wilbur Shepperson, *Restless Strangers* (Reno, University of Nevada, 1970), pp. 18–24.

45 See for example the diary of Mary McNair Matthews, *Ten Years in Nevada, or Life on the Pacific Coast* (Lincoln, University of Nebraska, 1985), pp. 249–263.

46 Paul Sui, "The Sojourner," *American Journal of Sociology 58* (1952), pp. 34–44.

47 See Randall E. Rohe, "After the Gold Rush: Chinese Mining in the Far West," *Montana: The Magazine of Western History* (Autumn 1982), pp. 2–19, for a survey of Chinese placer mining and miners.

48 Eugene Hattori, "Northern Paiutes on the Comstock," *Nevada State Museum Occasional Papers 2* (Carson City, 1975), p. 2.

49 *Ibid.*, table 3.

50 Elmer Rusco, *"Good Times Coming?": Black Nevadans in the Nineteenth Century* (Westport, Conn., Greenwood Press, 1975), p. 128.

DONALD CHAPUT

1 This account is based on reports in the *Mining & Scientific Press* (San Francisco), December 14, 1867, and the *Inyo Independent* July 16, 1870.

2 *Ibid.* (*Inyo Independent*), and W. A. Chalfont, *The Story of Inyo* (n.p., 1922), p. 249.

3 May 26, 1866, p. 326; November 30, 1866, p. 2.

4 Los Angeles *News*, December 3, 1867, p. 2.

5 *Mining & Scientific Press*, November 9, 1867, p. 295.

6 *Ibid.*

7 *Ibid.*, October 5, 1867, p. 214.

8   Archivo de Hidalgo del Parral, 1707a, Document 12, microfilm at the University of Arizona Library.

9   Quoted in *Mining & Scientific Press*, September 21, 1867, p. 182.

10   Virginia City *Trespass* and *Enterprise* quoted in *Mining & Scientific Press*, September 28, 1867.

11   *Mining & Scientific Press*, December 7, 1867, p. 358; December 14, 1867, pp. 382–3.

12   *Ibid.*, February 25, 1868, p. 102.

13   *Ibid.*, February 1, 1868, p. 78.

14   *Territorial Enterprise*, April 2, 1868, p. 3. Delavan's background appears in his obituary, *ibid.*, October 23, 1891.

15   Joseph Tasse, *Hoistoire des Canadiens de l'Ouest* (Montreal, 1878), II, pp. 250–60.

16   Remi Nadeau, *City Makers* (New York, 1948), p. 39; Robert C. Likes and Glenn R. Day, *From This Mountain: Cerro Gordo* (Bishop, [California], 1975), p. 10.

17   Obituaries in the Los Angeles *Evening Express*, March 8, 1888, and the Montreal *Gazette*, March 8, 1888.

18   *Southern California Quarterly*, VIII (1909–11), p. 231, for Santa Anita; the Los Angeles *Star*, April 6, 1861, for Azuza.

19   Likes and Day, *Cerro Gordo*, p. 10.

20   F. J. Hulaniski, *History of Contra Costa County* (Berkeley, 1917), pp. 432–33, and Alonzo Phelps, *Contemporary Biography of California's Representative Men* (San Francisco, 1882), pp. 159–60.

21   For example, the Cinco Señores had a San Francisco office; c.f. *Mining & Scientific Press*, October 19, 1863, p. 1. For quotation, c.f. Phelps, *California's Representative Men*, p. 160.

22   Phelps, *California's Representative Men*, p. 160.

23   Nadeau, *City Makers*, p. 32.

24   Los Angeles *News*, June 23, 1868, p. 3.

25   Nadeau, *City Makers*, pp. 34–35.

26   For the difficulties involved in mining at Cerro Gordo, see generally Nadeau, *City Makers*, and Likes and Day, *Cerro Gordo*.

27   Nadeau, *City Makers*, has a chapter on the *bandidos*. For the robbery of Beaudry, c.f. the Los Angeles *Star*, May 11 and 12, 1875.

28   *San Felipe Mining Co. v. M. W. Belshaw, et al.*, 49 Cal. 655, 655–58 (April, 1875). The legal tangle is also well covered in Nadeau, *City Makers*.

29   Nadeau, *City Makers*, pp. 41–42.

30   Quoted in the *Inyo Independent*, December 19, 1874.

31   Nadeau, *City Makers*, pp. 96–97. See also Harris Newmark, *Sixty Years in Southern California 1853–1913* (New York, 1916), pp. 386–88.

32   Rev. John Hemphill, *Memoir of Thomas H. Selby* (n.p., 1875) at the Huntington Library [San Marino, California] no. 36143. Hemphill was pastor of Calvary Presbyterian Church in San Francisco, Selby's congregation.

33   *Mining & Scientific Press*, May 4, 1867, p. 281; November 14, 1868, p. 312.

34 From metallurgical analysis performed in 1975 on ingot in collections of Natural History Museum of Los Angeles County, Museum Registrar File P.969.75.

35 *Eighth Annual Report* [of the] *State Mineralogist* (Sacramento, 1888), p. 228, reports that the population of Cerro Gordo in 1888 was between forty and fifty.

36 Adolph Knopf, *Mineral Resources of Inyo and White Mountains*, Bulletin 540, U.S. Geological Survey (1914), p. 99.

37 Obituaries in San Francisco *Chronicle*, April 30, 1898, and *Contra Costa County Gazette* (Martinez), April 30, 1898; Hulanski, *Contra Costa County*, pp. 432–33; Elise S. Benyo, *Antioch to the 'Twenties* (Antioch [California], 1972), at the Contra Costa County Library.

38 Obituaries in Montreal *Gazette*, March 8, 1888, and Los Angeles *Evening Express*, March 8, 1888. See also *Los Angeles City & County Directory 1866–1867*, pp. 99 and 192.

39 Donald Chaput, *French Canadian Contributions to Winning the American West* (Quebec, 1985), pp. 61–64.

GERALD W. R. WARD

1 Jervius Anderson, *Guns in American Life* (New York: Alfred A. Knopf, 1984), pp. 47–48.

2 Peter J. Bohan, *American Gold, 1700–1860* (New Haven: Yale University Art Gallery, 1963), pp. 6–8.

3 Much of the approach taken in this essay is based on my article " 'An Handsome Cupboard of Plate': The Role of Silver in American Life," in Barbara McLean Ward and Gerald W. R. Ward, eds., *Silver in American Life: Selections from the Mabel Brady Garvan and Other Collections at Yale University* (New York: The American Federation of Arts; Boston: David R. Godine, 1979), pp. 33–38, and on chapter 3 of my dissertation, "Silver and Society in Salem, Massachusetts, 1630–1820: A Case Study of the Consumer and the Craft" (Ph.D. dissertation, Boston University, 1984).

4 Neil McKendrick, John Brewer, and J. H. Plumb, *The Birth of a Consumer Society: The Commercialization of Eighteenth-Century England* (Bloomington: Indiana University Press, 1982), documents and discusses the "consumer revolution" in which wider ownership of small articles of silver played a part; also see Ward, "Silver and Society in Salem," chapter 1, for a case study of this phenomenon on the local level.

5 Elliot Evans and Paul Evans, "San Francisco Silver: Vanderslice & Company Flatware," *Spinning Wheel* 33, no. 10 (December 1977): 18–20.

6 Dorothy T. Rainwater and H. Ivan Rainwater, *American Silverplate* (Nashville: Thomas Nelson, Inc.; Hanover, Penn.: Everybodys Press, 1968), p. 38, figs. 13, 14.

7 Daniel Boorstin, *Image: A Guide to Pseudo-Events in America* (New York: Atheneum, 1962).

8 Benjamin White, *Silver: Its History and Romance* (London: Waterlow and Sons Limited, 1920), pp. 59–60. White's figures were based on estimates made by R. W. Raymond, Commissioner of Mining Statistics, and by the Director of the United States Mint.

9 Rodman Wilson Paul, *Mining Frontiers of the Far West, 1848–1880* (Albuquerque: University of New Mexico Press, 1974), p. 5 (map of silver deposits).

10 Charles H. Carpenter, Jr., with Mary Grace Carpenter, *Tiffany Silver* (New York: Dodd, Mead and Company, 1978), pp. 19–21.

11　See Paul Evans, "Shreve & Co. of San Francisco: Souvenir Spoons," *Spinning Wheel* 35, no. 4 (May 1979): 22–25, and Paul Evans, "Hammersmith & Field: San Francisco Souvenir Spoons," *Spinning Wheel* 35, no. 6 (September 1979): 26–28.

12　Quoted in Katharine Morrison McClinton, *Collecting American 19th Century Silver* (New York: Charles Scribner's Sons, 1968), p. 59.

13　The various levels of meaning conveyed by gifts are discussed by Marcel Mauss, *The Gift: Forms and Functions of Exchange in Archaic Societies*, trans. Ian Cunnison (Glencoe, Illinois: The Free Press, 1954). Mauss' (1872–1950) ideas on gift exchange have been little utilized by students of American silver. Robert B. St. George has called attention to them in relationship to the motives of donors of church silver in his essay "Artifacts of Regional Consciousness in the Connecticut River Valley, 1700–1780," in *The Great River: Art & Society of the Connecticut Valley, 1635–1820* (Hartford: Wadsworth Atheneum, 1985), p. 35.

14　Charles H. Carpenter, Jr., *Gorham Silver, 1831–1981* (New York: Dodd, Mead and Company, 1982), pp. 267–268.

15　See Gregor Norman-Wilcox, "American Silver at Los Angeles County Museum: The Collection of Mrs. John Emerson Marble," *The Connoisseur Year Book, 1956*, ed. L. G. G. Ramsey (London: The Connoisseur, 1956), pp. 62–70; Lorenz Eitner, *Catalogue [of] The Mrs. John Emerson Marble Collection of Early American Silver* (Stanford, Cal.: Stanford University Museum, 1976); Robert R. Wark, *British Silver in the Huntington Collection* (San Marino, Cal.: Huntington Library, 1978); William Ezelle Jones, *Monumental Silver: Selections from the Gilbert Collection* (Los Angeles: Los Angeles County Museum of Art, 1977).

16　Mihaly Csikszentmihalyi and Eugene Rochberg-Halton, *The Meaning of Things: Domestic Symbols and the Self* (Chicago: University of Chicago Press, 1981), chapter 3 and *passim*.